1976

University of St. Francis
GEN 621.38 D923

SO-BMV-505

3 0301 00022702 1

be kept

IRT EN DAYS

Communications
in Space

Other books by Orrin E. Dunlap, Jr.

DUNLAP'S RADIO MANUAL
THE STORY OF RADIO
ADVERTISING BY RADIO
RADIO IN ADVERTISING
THE OUTLOOK FOR TELEVISION
TALKING ON THE RADIO
MARCONI: THE MAN AND HIS WIRELESS
THE FUTURE OF TELEVISION
RADIO'S 100 MEN OF SCIENCE
RADAR: WHAT IT IS AND HOW IT WORKS
UNDERSTANDING TELEVISION
DUNLAP'S RADIO & TELEVISION ALMANAC

Communications in Space

FROM MARCONI TO MAN ON THE MOON

New and Expanded Edition

ORRIN E. DUNLAP, Jr.

LIBRARY
College of St. Francis
JOLIET, ILL.

HARPER & ROW, PUBLISHERS
New York, Evanston, and London

COMMUNICATIONS IN SPACE, NEW AND EXPANDED EDITION. Copyright © 1962, 1964, 1970 by Orrin E. Dunlap, Jr. All rights reserved. Printed in the United States of America. No part of this book may be used or reproduced in any manner whatsoever without written permission except in the case of brief quotations embodied in critical articles and reviews. For information address Harper & Row, Publishers, Inc., 49 East 33rd Street, New York, N.Y. 10016.

LIBRARY OF CONGRESS CATALOG CARD NUMBER: 75-108938

621.38
D923

To Louise

73180

To Loaoke

Contents

PREFACE   ix

  I. WIRELESS TELEGRAPHY   1
       Achievements of the Pioneers

  II. RADIO BROADCASTING   28
       The Miracle of Electronics

  III. RADIOPHOTO   55
       Pictures Across the Hemispheres

  IV. TELEVISION   63
       Sight Gives Radio a New Dimension

  V. RADAR   73
       A Science Based on Radio Echoes

  VI. SPACE AGE COMMUNICATIONS   86
       Astronauts Blazed the Trails

  VII. TRAIL-BLAZING THE LUNAR AGE   112
       Scientific, Economical, Political Aspects

  VIII. EPIC SIGNALS FROM THE MOON   136
       TV Showed "One Giant Leap for Mankind"

  IX. PIONEERING FOR PROGRESS   150
       Steps Toward a Satellite System

X.   COMMUNICATIONS SATELLITES   174
          "Switchboards" in the Sky

XI.   SATELLITES EXPLORING FOR SCIENCE   191
          Laboratories and Observatories in Space

XII.   ROBOT COMMUNICATORS   221
          From the Moon, Venus, and Mars

XIII.   TWINS OF SCIENCE: MASER AND LASER   246
          New Tools of Communications and Industry

XIV.   NEW IDEAS, NEW INVENTIONS   269
          Impact of the Space Age on Technology

XV.   ADVANCES IN THE SPECTRUM   286
          Innovations in Systems and Services

XVI.   EXPLORING THE UNIVERSE   295
          Probing the Cosmos, Planets, and Stars

XVII.   TESTS OF TIME   316
          Challenges Inventions Encounter to Survive

INDEX   331

Illustrations follow page 146.

Preface to this new and expanded edition

PROGRESS IN SPACE CONTINUES AT AN ACCELERATED PACE, AND increased activity is foreseen for the years ahead. When this book was first published, in 1962, it had eight chapters; this updated edition has seventeen required by subsequent events and inventions.

The first edition went to press after John Glenn orbited the earth in February 1962. Up to 1970, 24 American astronauts and 21 cosmonauts (one a woman) established historic records in orbital flight and in communications. Some of them made several journeys into space. This book records their triumphs as communicators. Herein is the story of the Mercury, Gemini, and Apollo projects—landing men on the moon—in which communications had a prodigious role—*the lifeline to the moon and back.*

This new edition brings the historic record up to date and looks into the future. It reports all of the orbital flights and accomplishments in communications, including the record-breaking globe-encircling network that tracked the astronauts and talked with them while they whirled around the world and the moon as millions on earth listened and watched. Aside from tracking, radar guides modules to rendezvous and dock, then to land on target. Computers too perform new roles not only in communications but in testing rocket engines, instruments, and systems. Computerized precision in navigation automatically leads to pinpoint splashdowns that TV cameras are nearby to see.

New laser-maser developments warrant a special chapter; it shows how the laser "family" has grown, explains how the laser works, and focuses on its great potential in communications and its applications in industrial and medical areas as well as in optics and research.

Indicative of how fast things happen in space—when this book first appeared, in 1962, no communications satellite had gone into orbit. Since that time Telstar, Relay, Syncom, Early Bird, and Intelsats opened a new era in international communications. The story of their success as scientific achievement and in relaying all sorts of information is fascinating reading. Plans for the future and activities of the Communications Satellite Corporation point to extension of the satellite system to help meet the increased demands for communications.

Many satellites, rockets, and unmanned spaceships have gone into space. Several in reconnaissance with Venus were heard by radio across more than 49 million miles, establishing new records in long-distance communication. But it was not long before those records were surpassed by satellites that sent back pictures of Mars 134 million miles away. Spacecrafts with TV cameras and special photographic equipment have relayed thousands of lunar pictures.

An ever-increasing number of instrumented satellites and solar observatories are probing space. The new knowledge they are collecting about the earth's atmosphere, the ionosphere, solar winds, magnetic fields, cosmic rays, the Van Allen radiation belt, and other mysteries of the universe constitutes an interesting chapter that sheds light on things man never suspected or merely surmised for centuries.

Communications itself is advancing on all fronts as new techniques and improvements in old concepts and instruments are put into service, for example, microwaves and troposcatter, while communication on beams of light is evolving. Many new products, materials, instruments and systems are emerging as a result of Space Age challenges.

Billions of dollars have been spent and billions more are earmarked for the conquest of space for peaceful purposes. Aerospace and electronics have become billion-dollar industries. In fact, economists declare that the national economy is hitched to the infinite and will be as far as they can see ahead. Years ago, radio put communications into the infinite now to become the life-line of manned space flight and the voice of exploring satellites that probe and whisper interplanetary secrets.

The first ten years of the Space Age made the universe instead of all the world the stage of communications. This book, therefore, in its aim "to hold, as 'twere the mirror up to nature," portrays the historic achievements of the past and present; it renews and expands its objective to show what has been done, what can and will be done in the future.

O. E. D., Jr.

Communications
in Space

1 Wireless Telegraphy

Let there be light:

—Genesis I. iii

"LET THERE BE LIGHT: AND THERE WAS LIGHT." AND THERE WOULD be radio too, for radio is akin to light.

Centuries would pass before the world was ready for such a system of communication as wireless telegraphy. Man first had to discover that a wondrous all-pervading medium existed, and then create instruments to set it in motion electrically, carrying messages, voices, music, and pictures. It was not a job for one man, or for one generation. In fact, the very idea seemed incredible.

Radio remained hidden in what scientists called "the ether"— defined as an imponderable, electric medium supposed to pervade all space as well as the interior of solid bodies; an invisible, odorless, tasteless substance was assumed to exist, through which light, heat, and radio waves could be transmitted.

MAXWELL'S THEORY

It was in 1867 that James Clerk Maxwell, physicist and mathematician at the University of Edinburgh, mathematically reasoned and theoretically outlined, in a famous equation, the action of "ether" waves. Inspired by an idea, gleaned from experiments, that the attraction or repulsion produced by electricity and magnetism were caused by some "action at a distance"—by an unseen medium in space—

Maxwell set out to find, mathematically at least, the missing link. As a result he identified what he called "the ether" believed to permeate the universe. He concluded that light and heat were electromagnetic undulations in the ether. And he was credited with having discovered "an elemental ocean in which the truth may yet be found."

Time alone would test the theory and reveal the truth. In any case, the theory helped to explain the mystery and to serve as a clue to what awaited mankind in space. Finally, however, the Maxwell theory was cast aside. Charles Proteus Steinmetz, in 1922, declared, "There are no ether waves; radio and light waves are merely properties of an alternating electromagnetic field of force which extends through space."[1]

The waves were not ethereal but electromagnetic. Apparently the world had been spinning in a vast unfathomed electromagnetic field since the beginning of time. The field was placid and at rest as far as man was concerned, until he built a wireless transmitter that caused it to vibrate. In the simplest terms, the action of the transmitter is like tapping a mold of jelly. Waves pass through it. The receiving set detects and translates the vibrations into sound and sight.

Maxwell's theory made it possible for teachers to compare the ether with a pond, in an effort to explain the action of wireless. The ether was pictured as a tranquil pool; toss in a stone and a series of ripples or waves are created, depending upon the size of the stone and the force with which it strikes the water. That stone is in effect the "transmitter." If tiny floating objects such as pieces of wood or cork float on the surface, they bob up and down in accordance (or in tune) with the waves; they are the "receivers."

Near the end of his life in 1879, Maxwell said that his only desire had been "to serve my own generation by the will of God." In truth he served far beyond his own time; his masterful treatise,[2] "Electricity and Magnetism," on the electrodynamic theory of light, speculated on the possibility of producing electromagnetic waves which would detach themselves from a source of origin—which, in other words, would be wireless.

Observing the 100th anniversary of Maxwell's birth, *The New York Times*[3] noted that British science in honoring him "both digs a

[1] Article by Steinmetz in *Popular Radio* July 1922.
[2] Presented to the Royal Society in 1864.
[3] October 5, 1931.

grave and erects a monument. The grave receives the remains of his theory of a luminiferous ether; the monument is to his mathematical genius. which ranks with that of Einstein. . . . Were Maxwell alive he would probably concede that his ether was no more real than 'the average man' of statisticians, or the equator of geographers—that it was necessary and convenient fiction without which the science of his day was helpless."

Said Einstein in an address at Nottingham University, "Looking back now we must ask why the ether as such was introduced? Why was it not called 'state of ether' or 'state of space'? The reason was that they had not realized the connection or lack of connection between geometry and space. Therefore, they felt constrained to add to space a variable brother, as it were, which would be a carrier for all electromagnetic phenomena. . . . It now appears that space will have to be regarded as a primary thing, with matter only derived from it, so to speak, as a secondary result. We have always regarded matter as a primary thing and space as a secondary result. Space now is turning around and eating up matter. Space is now having its revenge."[4]

HERTZ CREATED THE WAVES

Fascinated by the Maxwell theory, Heinrich Rudolph Hertz, a German physicist, took up the task of confirming it. As a result, he was the first to create, detect and measure electromagnetic waves. And he found a close kinship between these invisible waves and light; he observed that the velocity of the electromagnetic impulses was the same as that of light—186,000 miles a second.

The law of this electrical radiation, he discovered, was the same as the corresponding law of optics. He even demonstrated that electric waves could be reflected, refracted, and polarized like waves of light. And he went so far as to express belief that if the rate of oscillation could be increased a millionfold, the electromagnetic waves would actually "wash" upon the frontiers of light.

Hertz, in 1888, described his epochal experiments in a paper entitled, "Electromagnetic Waves in the Air and Their Reflection." By 1890, scientists declared that his discovery was working a revolution in their ideas about electricity.

[4] Lecture at the University College of Nottingham, June 6, 1930.

"Three years ago electromagnetic waves were nowhere; shortly afterward they were everywhere," exclaimed Sir Oliver Heaviside, English scientist and distinguished mathematician, in 1891. "The great gap between Hertzian waves and light has not yet been bridged. But I have no doubt that it will be by the discovery of improved methods of generating and observing very shortwaves."

Prophetic was his foresight. There would come a day when electron tubes would generate great power for shortwaves and substantiate electromagnetic observations that seemed fantastic at the turn of the century.

It was in 1902 that Sir Oliver presented his famous "guidance hypothesis" that depicted a conducting layer in the upper atmosphere. He noted that in transmission of radio signals along a pair of wires, the waves could be guided around a curved path so long as the path changed direction gradually.

This is how he explained it:[5]

Guidance is obviously a most important property of wires. There is something similar in wireless telegraphy. Sea water, though transparent to light, has quite enough conductivity to make it behave as a conductor for Hertzian waves, and the same is true in a more imperfect manner of the earth. Hence the waves accommodate themselves to the surface of the sea, in the same way as waves follow wires. The irregularities make confusion, no doubt, but the main waves are pulled around by the curvature of the earth, and do not jump off.

Then came that part of the hypothesis that linked his name with long-distance radio and with the electromagnetic aspects of space:

There is another consideration. There may possibly be a sufficiently conducting layer in the upper air. If so, the waves will, so to speak, catch on to it more or less. Then the guidance will be by the sea on one side and the upper layer on the other.

In effect, Heaviside pictured space between the earth and the conducting medium as a kind of wave-guide. As time went on the conducting layer in the upper atmosphere was popularly envisaged as a "ceiling" or "mirror" that reflects radio waves back to earth.

[5] Article by Heaviside in *Encyclopaedia Britannica,* 1902, and reprinted in "Electromagnetic Theory, Vol. III" under heading "Theory of Electric Telegraphy" published in 1912.

At the same time, Arthur E. Kennelly,[6] British mathematical physicist who came to America in 1887, working independently, was convinced that high in the earth's atmosphere a conducting layer of ionized air reflected radio waves, so that the earth's curvature was overcome. He did not assume the existence of a purely speculative conducting stratum but showed that it must exist at a height of about 50 miles because of rarefaction of the atmosphere with electrical conductivity several times as great as sea water.

Hertz, in his early experiments, had focused the waves with galvanized sheet acting as large mirrors, little realizing that in the ionosphere Nature had a "mirror" that would reflect them on a world-wide scale. He also refracted the waves with prisms made of coal-tar pitch. And he noticed that a tinfoil screen or a person in the path of the waves, cast a "shadow." He wrote:[7] "It [the wave] passes through a wooden partition or door, and it is not without astonishment that one sees sparks appear inside a closed room."

Yet Hertz showed no prescience about the possible use of the waves; it was said that he even argued that they could not be of any practical use.

Nevertheless, Heinrich Hertz was the founder of a new epoch in experimental physics. His experiments with cathode rays, to determine whether they were electromagnetic, or a beam of electrically charged particles, fell just short of crossing into the world of the electron and the atom. He subjected cathode rays in a vacuum to a strong electric field. If the waves were charged particles, the field should deflect the beam; if they were waves they would ignore the field. Hertz was misled to believe that they were unaffected by the field, because the vacuum was inadequate; residual gas had shielded the waves from the field.

But as long as there is electromagnetic communication in space, the Hertzian waves will vibrate in tribute to him, while history records that on the unerring accuracy of this brilliant physicist rests

[6] Professor Kennelly and Sir Oliver Heaviside were recognized as co-discovers of "the radio mirror," and it became known as the Kennelly-Heaviside layer, or surface. Kennelly served as an assistant to Edison and in 1902 was appointed Professor of Electrical Engineering at Harvard University.

[7] "Electromagnetic Waves and Their Reflection," published by Hertz in May 2, 1888, as one of a series of papers addressed to the Berlin Academy of Sciences between November 1887 and December 1889.

the foundation of wireless, radio broadcasting, radar, television, and the technology of satellite signaling.

At the age of thirty-seven, Hertz passed away on January 1, 1894, but not before he had passed the electromagnetic torch to Guglielmo Marconi and other men of science, whom he inspired to look into the new regions of the electromagnetic spectrum which he had so successfully explored.

MARCONI ENTERS THE SCENE

While vacationing in the Alps, *Guglielmo Marconi,* a young Italian in his teens, picked up an electrical journal, *Wiedemann's Annalen,* which told how Hertz had created and radiated electromagnetic waves, some measured in meters and others in centimeters—ultrashort! For Marconi it was the germ of an idea: Why not use the waves for signaling? He cut short his vacation and rushed back to his home at Pontecchio, near Bologna, Italy, to try what must have seemed fantastic.

Wireless at that stage in 1894 was an electrical jigsaw puzzle. Marconi put together an induction coil, the Hertz wave emitter, the Righi spark-gap, a telegraph key, batteries and a Branly coherer.[8] And he hoped that the combination might enable him to send and receive signals across his father's estate.

The transmitter consisted of an aerial circuit with one end connected to an elevated aerial and the other end to the ground. The circuit contained a spark gap the terminals of which were connected to the secondary terminals of the transformer, or induction coil. The primary of this coil was connected to the source of current (batteries) and a telegraph key for signaling with dots and dashes. The low-frequency current was caused to discharge through the spark gap, producing high-frequency oscillations which were radiated by the aerial.

---

[8] The coherer was invented by Edouard Branly, French physicist, who observed that some metallic powders or particles in a glass tube were affected in their electrical conductivity by electromagnetic waves. When signals were impinged on metallic dust particles they cohered; a tiny hammer, like one used on a doorbell, tapped against the tube to decohere them, stopping the current from a local battery. Each successive impulse from the antenna produced the same phenomena of coherence and decoherence; thus the name coherer.

The receiver similarly featured an antenna circuit between an elevated plate and the ground, and into that circuit was connected the coherer (detector). The transmitter and receiver were constructed to have the same resonant frequency; this was accomplished by careful determination of the size and height of the aerial plates.

Marconi succeeded where others had failed by making use of the so-called ground return; i.e., both the transmitter and receiver had connections with the ground. He heard cricket-like sounds across three-quarters of a mile—and that marked the beginning of wireless telegraphy.

Marconi's Story   Marconi narrated the advent of wireless as a matter-of-fact story:[9]

"From a boy I was always interested in physics and in electric phenomena generally, and in the summer of 1894 I read of the experiments and results of Hertz in Germany. I was also acquainted with the works of Lord Kelvin and with the theoretical doctrines of Clerk Maxwell. I experimented with electrical waves, as I considered that line of research very interesting. During these tests or experiments I thought that these waves, if produced in a somewhat different manner—that is, if they could be made more powerful, and if receivers could be made more reliable, would be applicable for telegraphing across space to great distances.

"In 1894 and the beginning of 1895, I constructed apparatus which was practically the same as the original Hertz apparatus for transmitting these waves," continued Marconi. "And I made a receiver which contained a detector which manifested or revealed the presence of the waves at the receiving end. I had always in mind telegraphic transmission and for this reason I had a Morse signaling key at the sending end and some kind of a detector which could be used telegraphically at the receiving end."

Describing the original set-up, Marconi said the transmitter, which consisted of the induction coil and spark gap, was at one side of a table. To the spheres of the spark gap were attached two metallic plates. The primary of the induction coil was energized from a battery, the current of which went through the telegraph key. When

---

[9] Testimony in the injunction suit of the Marconi Wireless Telegraph Company of America against the National Electric Signaling Company in the United States District Court, Brooklyn, N.Y., 1913.

the key was pressed the current passed through the coil and was transformed into high-tension electricity which caused sparks to jump across the gap. And Marconi explained that according to principles then well-known and promulgated by Hertz and others, when the sparks occurred the spark gap and plates radiated electric waves into space.

Up to this time Marconi had no tuner. He explained how he got around it:

"A straight wire has both capacity and inductance, and I could change the inductance by cutting off part of the wire. That was the only way I adjusted the period (tuning) of my early apparatus. At that time I did not use any coils of wire. My antenna in 1896, varied from six to thirty or forty feet. If I wanted to change the period (tuning) I let the antenna down and cut it off at the bottom. I first used an inductance coil (tuner) in the antenna about the end of 1898."

Then he went on to tell how the waves were received:

"These waves travel across the distance, which in this case was simply from one end of a table to the other. The waves produce induced currents between the two resonating or collecting plates (antenna) of the receiver. An improved Branly tube (coherer) was connected in the circuit with a battery and telegraph instrument. The tube was a non-restoring coherer, that is, after being influenced by electric waves it would go on conducting. Without that action it would be impossible to reproduce dots and dashes of the Morse code. To restore it and keep it in its sensitive condition, the same current which actuated the telegraph instrument acted upon a tapper which tapped the tube (coherer) and restored its sensitiveness after receipt of the impulses. In that manner it was possible to transmit the signs of the Morse code over a distance of three or four feet!"

AN EPOCHAL DISCOVERY

The coherer was a primitive device. If wireless telegraphy was to be a commercial success a more sensitive and stable detector would have to be found, as well as amplifiers and oscillators. An invention that would accomplish those three steps seemed impossible.

Nevertheless, while Marconi was doing his best with the coherer, an epochal discovery was made, destined to revolutionize the entire science of communications, including the telephone. Virtually unnoticed, as far as wireless was concerned, was the discovery of the electron by J. J. Thomson, English physicist. In 1897, he demonstrated the true character of the electron as the smallest particle of the electrical structure of the atom.[10] He did not isolate, or see it, but he verified that things much smaller than the atom existed, and that these entities had the same mass and negative electric charge regardless of the kind of metal from which they were produced. The electron was the key to a new age in electrophysics and, unsuspected by the veterans of wireless, it was a key to a vast new world of radio. First, however, some new tool had to be found to generate electrons, liberate them, amplify and control them, and put them to work, thereby creating the science of electronics.

THE SCENE SHIFTS TO ENGLAND

Up to the end of the nineties no one knew how to harness electrons, so Marconi—still depending upon the coherer—went to England in 1896, for it appeared that the greatest opportunities of wireless were linked with the sea, the waves of which, at that time, Britannia ruled.

"I first offered wireless to Italy," he explained, "but it was suggested, since wireless was allied to the sea, it might be best that I go to England, where there was greater shipping activity, and, of course, that was the logical place from which to attempt transatlantic signaling. Also my mother's relatives in England were helpful to me. I carried a letter of introduction to Sir William Preece.[11] Mind you, Italy did not say the invention was worthless, but wireless in those days seemed to hold promise for the sea, so off to London I went."

Utilizing ultra-shortwaves and reflectors, Marconi projected beamlike waves to demonstrate across 100 yards, and in the summer of 1896 at Salisbury Plain, England, wireless leaped nine miles. That was the year he applied for his original and basic patent (No.

[10] Thomson first called the tiny particles "corpuscles," the existence of which he announced at the Royal Institute on April 20, 1897, and published in the *Electrician* on May 21, 1897. He won the Nobel Prize for physics in 1906 and was knighted in 1908.

[11] Engineer-in-Chief of the British Post Office.

12039), which described the use of a transmitter and the coherer connected to the earth and elevated antenna; the equivalent American patent was issued in 1897.

France wondered if wireless could signal across the English Channel. Marconi went to France to prove it; on March 27, 1899, he flashed messages across 30 miles from Wimereaux to the cliffs of Dover. That feat, he described as "the most important thing I did"— up to that time. Then the French, as well as the United States, British, and Italian Navies, all became vitally interested in wireless for warships and shore stations.

Once Marconi had succeeded in putting the signals into the air, he was challenged to overcome what was considered a weakness in wireless—how could two or more stations operate at the same time without overlapping and interfering with each other? Marconi found the answer. He invented a tuner to eliminate interference and clear the way for any number of stations to operate. His famous British patent No. 7777 on 4-circuit tuning was granted in 1900. So by the turn of the century, wireless under the impetus of its inventor showed promise of a great future, for this serious-minded Italian youth had the money, courage, patience, initiative, perseverance, intuition and other assets needed to carry his invention into commercial use. He didn't claim to be a scientist. Modestly he explained that he had observed certain facts and developed certain devices to meet them.

Sir William Preece, himself deeply interested as an experimenter in signaling through space without wires, summed it up: "While I cannot say that Mr. Marconi has found anything absolutely new it must be remembered that Columbus did not invent the egg. He showed how to make it stand on end."[12]

TRANSATLANTIC WIRELESS

There was one big hurdle Marconi wanted wireless to leap, and once that was accomplished, all other goals would be within reach. *Transatlantic*—that was the challenge, and the big prize Marconi cherished. He built a powerful station at Poldhu on the southwest tip of England. When it was ready for the big test he went to St. John's,

[12] Interview in *The New York World*, August 8, 1897.

Newfoundland, where he and two assistants, George S. Kemp and G. W. Paget, installed receiving apparatus. December 12, 1901, became a historic date in the annals of communication for on that day Marconi heard the letter "S"—three dots in the Morse code—flashed from Poldhu to Newfoundland, 1,800 miles!

Marconi mustered everything he could get in the way of detectors, and even the best were none too sensitive. His coherer had a tapper and a coil; he also tried a carbon microphonic detector. Another, which he rated as the most successful, consisted of an imperfect contact between two pieces of carbon, the contact being made by a globule of mercury as previously used by the Italian Navy.

The copper wire antenna was held aloft by a kite that gyrated from four hundred to six hundred feet in the wintry wind over St. John's. In the public mind it was a miracle that a wireless signal from across the ocean could hit such a thread-like target.

It was such an incredible achievement that it is no wonder scientists as well as laymen just couldn't believe it. There were no end of doubting Thomases, but Thomas A. Edison was not among them.

"If Marconi says so, it's true," he declared. "I would like to meet that young man who has had the monumental audacity to attempt and succeed in jumping an electric wave across the Atlantic!"[13]

Sir Oliver Lodge, a pioneer in experimenting with electromagnetic waves as a means of signaling, said, "Marconi's creation, like that of the poet who gathers the words of other men in a perfect lyric, was none the less brilliant and original. . . . It constituted an epoch in human history, on its physical side, and was an astonishing and remarkable feat."[14]

"Remarkable is right," exclaimed Edwin H. Armstrong, a veteran of wireless, thirty years later. "It puzzles me how Marconi did it with the instruments then available. I understand the wavelength of Poldhu was about 250 meters. I have always wondered what stroke of luck, or freak atmospheric conditions, would enable that wave to span the sea especially at noontime when the sun is high. We know

[13] Message in tribute to Marconi read at dinner of the Institute of Electrical Engineers, January 13, 1902, commemorating the first transatlantic wireless signal.

[14] Statement in tribute to Marconi on success of first transatlantic wireless, December, 1901.

the sun absorbs about 70 per cent of the strength of wireless waves. That accounted for long-distance records being established at night, while the daytime range was limited. At least that was true until shortwaves were harnessed and the electron tube made receiving apparatus a thousandfold more sensitive and transmitters were vastly more powerful."

Under the circumstances, Armstrong agreed that if a long wave had been used, it might have led to success. He declared that Marconi's claim to the invention of wireless was beyond challenge, and it would be interesting to know exactly what wavelength he used for the transatlantic signal, whether 250, 1,000, or 2,000 meters.

No one would know better than Sir Ambrose Fleming what wavelength was flashed from Poldhu for as Scientific Advisor of the Marconi Wireless Telegraph Company in 1899, he played an important role in the design and installation of the 25-kilowatt station at Poldhu.

Asked to clarify the matter, Sir Ambrose replied:[15]

"The wavelength of the electric waves sent out from the Poldhu Marconi station in 1901 was not measured because I did not invent my cymometer or wavemeter until October 1904. The height of the original aerial (1901) was 200 feet but there was a coil of a transformer or 'jigger' as we called it, in series with it. My estimate was that the original wavelength must have been not less than about 3,000 feet (approximately 1,000 meters) but it was considerably lengthened later on.

"I knew at that time that the diffraction or bending of the rays around the earth would be increased by increasing the wavelength and after the first success I was continually urging Marconi to lengthen the wavelength, and that was done when commercial transmission began. And I remember I designed special cymometers to measure up to 20,000 feet (approximately 7,000 meters) or so."

Marconi did not stop with the transatlantic triumph; he was only twenty-seven years old and a future full of challenges to develop wireless to the utmost fired his youth. When the S. S. "Philadelphia" docked in New York in March 1902, he came down the gangplank and handed news reporters yards of "telegraph" tape, dotted and

[15] Letter to the author, April 16, 1935.

dashed with hundreds of signals automatically recorded from Poldhu while enroute across the ocean.

"Now," exclaimed Marconi, "will anyone say I was mistaken in Newfoundland!"

TESTED BY DISASTER AT SEA

There came a day when wireless was tested in the crucible of disaster—on January 23, 1909—when the passenger-laden S. S. "Republic," outbound from New York on a sunny Mediterranean cruise, collided with the Italian freighter "Florida," in a dense fog 26 miles southwest of Nantucket Lightship.

The CQD (distress call at that time) was flashed by Jack Binns, wireless operator and hero of the hour on the foundering "Republic." Answering his call, rescue ships rushed from all directions. The "Republic" went down, while the "Florida" with her bow cut away, slowly crawled into port. Because of wireless a great loss of life was averted—the "Florida" lost four seamen; the "Republic," two passengers out of 461.

Tragedy again struck with violence on the night of April 14, 1912, summoning everything that the Marconi wireless had to offer. Out of the night air, Charles B. Ellsworth, a seventeen-year-old Marconi wireless operator at Station MCE, Point Riche, Cape Race, Newfoundland, was startled by an unbelievable message from the majestic S. S. "Titantic" on her maiden voyage:

CQD, SOS from MGY (call of the "Titanic"). We've struck a berg. Sinking fast. Come to our assistance. Position, latitude 41.46 north, longitude 50.14 west, MGY.

Rescue ships went full-steam to the scene of disaster—the "Carpathia," "Olympic," "Baltic," "Virginian," "California," "Frankfurt," "Parisian," and others. The "Carpathia," 58 miles away, was nearest and reached the tragic spot at dawn only to find lifeboats and rafts among the wreckage, where the big ship had gone down at 2:20 A.M., several hundred miles off the Grand Banks of Newfoundland. The dead numbered 1,517; wireless had saved 712 lives. The "Carpathia" picked up the survivors and headed for New York.

"I was on the dog-watch 1 A.M. to 5 A.M. AST," said Ellsworth.[16] "At 1:34 A.M., I intercepted the CQD distress call from the 'Titanic.' I couldn't believe it and thought someone on the station was playing a trick on me. I looked into the bedrooms and everyone was asleep. Then I ran and woke the operator-in-charge. I'll never forget him standing there and telling me what would happen if I were mistaken. He came into the wireless room and heard the CQD being flashed by operator Jack Phillips. I knew then it was no hoax. We went sleepless for days, handling messages from the various rescue ships."

Forty-nine years later Ellsworth said, "I can still hear poor Phillips saying '250 miles SW Cape Race.' When he lost power from the engine room he went to his battery-operated 10-inch spark coil on the topside and we read him Okey. Whether he shifted to the coil before the engine room flooded I do not know, but he used the main 1.5-kilowatt transmitter, equipped with the first 500-cycle rotary spark that ever sailed out of Southampton, for at least a half hour after the first distress call.

"The 'Titanic' was violating all rules of the transatlantic trade," recalled Ellsworth, "when she took the northern great circle course, which is 135 miles shorter, to break the speed record. Liners never used that route until the middle of June or first of July depending on the iceberg reports. The bergs come down the east coast of New-foundland, in what fishermen call the inshore current that hugs the coast 10 to 30 miles east. They move down to the Grand Banks, where in some places the water is only twelve fathoms deep. Some bergs just ground right there and stay—and this shallow water is on the southern edge of the Banks, about 250 miles southwest of Cape Race. . . . Some stories reported the 'Titanic' 800 miles east of Cape Race, but that was a reporter's dream. We would never have heard the distress call had the ship been 800 miles east of the Cape."

Wireless in those days, even on such a luxury liner as the "Titanic," was not powerful enough to reach from mid-ocean to Boston, New York, and other east coast cities. The "Olympic's" range was esti-mated at 500 miles, and the "Carpathia's" much less.

Jack Binns at this time was a reporter on *The New York American,* and was assigned to the "Titanic" story.

[16] Told to the author at reunion of U.S. Navy radio operators of station NBD, Otter Cliffs, Maine, on August 19, 1961.

"Hearst hired a seagoing tug, the 'Scully,' to put to sea and meet the 'Carpathia,' " recalled Binns.[17] "I went to the Marconi Company and asked if they would equip the tug with wireless. They agreed if I would go aboard the 'Carpathia' and relieve Harold Cottam, the Marconi operator on board because he was fatigued—no sleep since he picked up the 'Titanic's' distress call.

"The wireless was quickly installed and at the same time I noticed that a lot of tin cans and butterfly nets were being put on board. I asked why such a cargo, and was told that the editors figured the 'Carpathia's' wireless could not be used under such circumstances for news dispatches. Therefore, the cans could be thrown up on the decks to enable passengers to write messages or news and throw them into the water; the butterfly nets would be used to scoop them up!

"Well, the 'Scully' took a position near 'Nantucket Lightship,' " said Binns. "Early that morning I heard the 'Carpathia' and she had already passed Nantucket; we missed her because she passed about 10 miles from the 'Lightship.' The 'Scully' was supposed to make 25 knots but could only make 6, so we never caught up to the 'Carpathia.'

"I copied the names of survivors as did the operators at Cape Race and other land stations as well as wireless amateurs. Dave Sarnoff was operator at the Marconi station atop Wanamaker's store in New York, which was designated as the center for reception from the 'Carpathia.' President Taft ordered all other stations to be silent so that messages from the 'Carpathia' would not suffer interference."

Marconi was acclaimed the world over for the invisible lifelines he had thrown out across the sea. He was in New York when the "Carpathia" docked, and was one of the few permitted to go aboard. Bowed in grief as he came down the gangplank he said, "It is worthwhile to have lived to make it possible for these people to have been saved."

AMONG THE PIONEERS

Despite the international acclaim, Marconi was the first to recognize that he was not alone in wireless; he was fully aware that an

[17] Told to the author at luncheon on January 23, 1959, the 50th anniversary of the S. S. "Republic" disaster.

inventor seldom covers his entire field and rarely wins by playing a lone hand. But those who sought to dim his glory by claims of predating him had no proof that they had previously accomplished what he had achieved.

Of Professor Alexander Stepanovitch Popoff, who claimed priority in wireless, Marconi said:[18] "Popoff's apparatus, prior to my patent, was a receiver for recording atmospheric electricity (static) and the effects of thunderstorms and lightning. Well, unless something in nature sent him impulses—and those impulses never come in the form of intelligent messages—I do not see where he was to receive signals from. I do not think his receiver was capable of taking Morse signals. It is easy enough to record the effect of lightning, but it takes much more accuracy to correctly interpret the signs of the Morse telegraph."

True there were others in the field, some of whom made outstanding contributions to the advance of electromagnetic waves, and they helped to pave the way to wireless. Among those who revealed through experiments that they had been on the threshold of signaling through space were: Thomas A. Edison, Sir Oliver Lodge, Sir William Preece, Reginald A. Fessenden, Professor Augusto Righi, Sir Ambrose Fleming, Sir Oliver Heaviside, Lord Kelvin, Edouard Branly, Sir William Crookes, Nikola Tesla, Professor George Pierce, Professor Amos E. Dolbear, Mahlon Loomis, Professor Adolphus Slaby, and others.

Marconi, mindful of their contributions, gave them full credit. Experience taught him that separate, unassembled and uncoordinated inventions must be brought together if the public is to have the benefit of each inventor's achievement. Putting them all together to operate in unison could be a long arduous task that called for further invention and ingenuity to harness them into a service or system.

SCIENCE AND THE LAWS OF NATURE

Observing that the discoveries of science are the discoveries of the laws of nature, and, like nature, do not go by leaps, Justice Felix

[18] Testimony in the injunction suit of the Marconi Wireless Telegraph Company of America against the National Electric Signaling Company in the United States District Court, Brooklyn, N.Y., 1913.

Frankfurter, of the U.S. Supreme Court, called attention to the fact that even Newton, Einstein, Harvey, and Darwin, built on the past and on their predecessors:[19]

Seldom indeed has a great discoverer or inventor wandered lonely as a cloud. Great inventions have always been parts of an evolution, the culmination at a particular moment of an antecedent process. So true is this that the history of thought records striking coincidental discoveries—showing that the new insight first declared to the world by a particular individual was "in the air" and ripe for discovery and disclosure.

The real question is how significant a jump is the new disclosure from the old knowledge. Reconstruction by hindsight, making obvious something that was not at all obvious to superior minds until someone pointed it out,—this is too often a tempting exercise for astute minds. The result is to remove the opportunity of obtaining what Congress has seen fit to make available.

The inescapable fact is that Marconi in his basic patent hit upon something that had eluded the best brains of the time working on the problem of wireless communication—Clerk Maxwell and Sir Oliver Lodge and Nikola Tesla. Genius is a word that ought to be reserved for the rarest of gifts. I am not qualified to say whether Marconi was a genius. Certainly the great eminence of Clerk Maxwell and Sir Oliver Lodge and Nikola Tesla in the field in which Marconi was working is not questioned. They were, I suppose, men of genius. The fact is that they did not have the "flash" that begot the idea in Marconi which he gave to the world through the invention embodying the idea.

Timeliness played a vital role for Marconi as the perfectionist who provided the master touch. His was an indefatigable interest in the science of communications, and he was endowed with an indomitable determination to develop his invention to the fullest extent. Destiny called him upon the scene at the opportune time when the world was ready for wireless and instruments were available for assembly to make it possible. As the saying goes, "nothing is stronger than an idea whose time has come."

"From the very beginning," he once declared,[20] "I learned it is unwise to put any limitations on the range of electric waves. Long experience has taught me not always to believe in the limitations indicated by purely theoretical considerations or calculations. These —as we well know—are often based on insufficient knowledge of all

---

[19] *Marconi Wireless Telegraph Company of America* vs. *United States,* June 21, 1943.
[20] Lecture at Royal Institution of Great Britain, December 2, 1932.

the relevant factors. I believe, in spite of adverse forecasts, in trying new lines of research, however unpromising they may seem at first."

Applying that doctrine along with electronics to the advance of communications he turned his attention to shortwaves, microwaves and radiotelephony, as he roamed the seas conducting experiments aboard his floating laboratory, the yacht "Elettra."

Suddenly, on the morning of July 20, 1937, Marconi's "race of existence" with wireless came to an end. An immense achievement had made his name immortal as he stood at the end of life's road with the infinite vista of electronics stretched before him. He had drawn the most distant places and many forgotten lives into the orbit of civilization.

THE MAGIC OF RADIO CONTROL

While Marconi was developing wireless for communication other experimenters, convinced that it was too powerful a force to be limited as a carrier of messages, began to test new applications. For instance, if radio could spread across the world, its beams should be able to reach out like fingers on a long arm and actuate control apparatus on ships, planes and even torpedoes at any distance.

John Hays Hammond, Jr., son of a distinguished mining engineer, was captivated by the idea after reading how Nikola Tesla in 1898 had fitfully controlled a model boat in a tank by wireless. Before graduation from Yale in 1910, young Hammond predicted that some day he would control a moving body at a distance by the sound of his voice. He established the Hammond Research Corporation at Gloucester, Massachusetts, in 1911, and his first invention was a little box on wheels, which he called an electric dog, that moved as he directed by radio or even by beams of light from a flashlight or lantern.

This fantastic dog was described as "a melancholy creature with eyes of bulging glass as large as saucers. His nose is a long thin strip of brownish reddish board. His body is an oblong mahogany box containing an electric motor, storage battery, two selenium cells, two electric relays and two solenoid magnets. The dog has no tail except an electric switch, and he runs on three brass wheels, two in front, one behind. When the motor inside the dog is started he will do some extraordinary things; he will even follow you at your radio command."

This scientific dog led to hundreds of patents pertaining to radio controls. From a 360-foot tower on the cliff at Gloucester, Hammond remotely controlled a 40-foot boat, "Radio," "with not a soul aboard, her lights blazing, her engines pounding, shooting out to sea at night with the speed of a railroad train." By manipulating the controls atop the lookout station he could command the 33-knot boat to zigzag, stop, go ahead, or turn around, full-speed or half-speed; rudder, engine speed, and searchlight all responded to radio control.

Said a newsman who watched the experiment,[21] "From his tower by touching a key, Hammond can send his boat out and back along an eight-mile course filled with rocks and shoals and harbor craft, at the speed of the swiftest cruiser. He can aim it at a mark three miles away and strike with precision every time. From this boat, running by wireless on the water, to a similar boat running under the water, is but a step. When that step is taken we shall have the ultimate torpedo for which the navies of the world are waiting."

Hammond realized that success of the crewless boat and the maneuvers of the electric dog formed the basis of "homing," the ability of a torpedo or missile to seek out its target. He turned his attention to radio-controlled, or radiodynamic torpedoes—sleek weapons that could be shot into the sea and be directed to hit a bull's-eye no matter what the target. To prove it, at the Newport Torpedo Station he developed the control of standard torpedoes running at a depth of twelve feet at a speed of 40 knots.

As early as 1912, Hammond established the principles of radio-guided missile control, and two years later demonstrated with experimental application the radio control of waterborne craft. He had a new boat, the "Natalia," which throughout a round trip from Gloucester to Boston was the first to utilize a gyroscope in an automatic pilot system to relieve the helmsman in holding a course. And the "Natalia" operated in and out of Gloucester harbor with the automatic pilot changed not by the helmsman but by radio-control signals from a transmitter on shore.

[21] Cleveland Moffett, *McClure's Magazine,* March, 1914.

"That gol-derned feller," exclaimed a Gloucester fisherman, "if he saw a big liner comin' on the rocks in a storm, he'd just fetch hold of her by wireless and turn her right plumb around and steer her out to sea again, by gosh!"[22]

All this was the forerunner of radio-controlled pilotless planes that would take off, fly and land completely under control and guidance from the ground. Big ships as well as small boats also could be controlled by radio as Hammond demonstrated in 1920, when he installed equipment on the battleship "Iowa" for tests off the Virginia Capes. Eighty dummy concrete "bombs" were dropped from planes as the defenseless "Iowa" was maneuvered in a zigzag course by radio control from the U.S.S. "Ohio," and thus avoided theoretical destruction.

From that day, many of Hammond's inventions were secret for security reasons. Only such developments as a television system for guiding aircraft to a safe landing in fog or storm, and any number of patents pertaining to radio instruments and circuits, including basic developments in intermediate frequency circuitry, which is a part of all radio and television systems, were made known. Evidence that his creative and inventive activity continued across the years is found in the fact that he owned nearly 700 patents.

It is a matter of record that during World War II, Hammond's radio-controlled glider bombs were released from high-flying American planes and destroyed the railroad yards at Cologne; the Azon bomb with Hammond controls pin-pointed bridges and enabled many bombers to remain untouched by anti-aircraft fire.

Well may one surmise as a modern missile, or rocket, races down range under perfect electronic control, or a spaceship lands "on the button," that Hammond's ingenuity and vision contributed to the success. Ever since the day of the electric dog he devoted his life to the development of radio controls extending all the way from boats and torpedoes to vehicles launched into space. In the annals of electronic communications John Hays Hammond, Jr., may well be registered as "the father of radio-guided missiles," for many developments in his laboratory established basic principles used in modern airborne guided missiles.

[22] *Ibid.*

ROLE OF THE AMATEURS

As Marconi's wireless and deForest's radiophone inspired many American boys, so too Hammond's radio-controlled boats and torpedoes awakened new interest in radio science. The romance of wireless, intensified by the dramatic news of its performance at the time of the "Titanic" disaster, had fired the imagination of youth and attracted school boys to wireless as a hobby.

No saga of wireless would be complete without the record of the amateurs, or "hams," as they call themselves. Their accomplishments are outstanding proof of the value of youth with an interest and aptitude in science. Indeed, radio progress and the amateurs are inseparable. They were a backbone of wireless and radiotelephony as evidenced by their discoveries and inventions and by the long list of notables in science, engineering and industry who came from their ranks. Self-trained at their home-built stations, the amateurs answered the call to the colors in two World Wars in which they manned Army and Navy stations ashore and afloat with impressive efficiency.

Before government licensing of amateur operators and stations was instituted in 1912 the amateurs could operate on any wavelength they chose and could select their own call letters. Long distance was their goal. To prove their efforts successful, Paul Godley, an ardent amateur from New Jersey, went to Ardrossan, Scotland, in December 1921, and picked up signals from twenty-seven amateurs in the United States as they transmitted across the Atlantic on power outputs ranging from 50 to 1,000 watts.[23]

As commercial activity expanded, the amateurs were assigned the 200-meter waveband which they found ineffective and condemned as "the graveyard of wireless." Confident, however, that there must be a more fertile place for them in the spectrum they began to investigate shortwaves—100 meters and shorter.

Proof that they were on the right track was found in November,

[23] The first message was received from station 1BCG, built by the Radio Club of America at Old Greenwich, Conn., operating on 230 meters, input 990 watts.

1923, when the first amateur two-way transatlantic communication was established by two enthusiasts in Connecticut—Fred H. Schnell at West Hartford and John L. Reinartz at South Manchester—who communicated with Leon Deloy at Nice, France, on a wave of about 100 meters. Encouraged by this achievement and aided by electron tubes the amateurs were continually establishing new long distance records, and commercial engineers, amazed at the results, also intensified their efforts in the shortwave spectrum.

During World War II radio amateur stations again were closed and many of their owners joined the service. Finally, when the wartime ban was lifted in 1945, sixty thousand amateur stations were ready to resume operation after four and one-half years of silence.[24] By this time the radiophone was a big attraction and many amateurs took to microphones instead of radio keys as the voice gained the glamour long possessed by the dots and dashes.

Always at the forefront, continually alert for anything new, the amateur experimenter has been and continues to be a trail blazer in communications. As early as 1916 two amateurs, George C. Cannon and Charles V. Logwood, broadcast music at New Rochelle, N.Y., from 9 P.M. to 10 P.M., daily except Sunday. And true to amateur traditions, in February 1960, Raphael Soifer in New York and Perry I. Kline in Bethesda, Maryland, 200 miles apart, established the first amateur two-way space communication by bouncing radio code signals off an artificial satellite believed to have been either Explorer 7 or Sputnik 3 in transit over the Atlantic seaboard.

WIRELESS BECAME RADIO

As time went on, both in amateur and commercial circles, the word "radio" gradually took the place of "wireless." Engineers and particularly the U.S. Navy felt that "radio" was more appropriate to distinguish the electromagnetic radiations through space from earlier conduction and induction systems which in effect were wireless too. Strictly speaking, "radio" includes electromagnetic wave signaling and nothing else. In the general public mind "wireless" signified dot-and-dash telegraphy and "radio" related to the radiophone and

---

[24] By 1950, there were 84,000 amateur stations in the United States and in 1970 the number approached 275,000.

broadcasting. After about 1912, "radio" became the more generally used term.

Under that classification the radio art, keeping pace with electronics and automation, continued to grow and perform new services on a world-wide scale undreamed of by the pioneers of wireless. No distance was too great for radio to span. It brought nations within a whisper of each other; television would put them within a wink.

RADIO IN AVIATION

As is often true, one industry leads on to another. As mariners encouraged development of wireless, aviation called upon radio to adapt itself to flight. The pioneer airmen shunned radio because it was too heavy, too bulky to qualify for wings. Space in the cockpits was limited, every ounce of weight counted, so the flyers reluctantly were forced to leave radio on the ground and take the risk of being in trouble without it.

When John Alcock and Arthur Whitten Brown took off from Newfoundland in June 1919, on the first non-stop transatlantic flight, they carried no radio transmitter. But when they landed in the mud alongside the towers of the big wireless station at Clifden, Ireland, upon which they had predicted "we will hang our hats," it was reported that a radio compass had guided them, using the waves from Clifden as a beam.

NC-Flying Boats and Their Successors   When the U.S. Navy sent three NC-flying boats on a transatlantic mission in May 1919, part of the purpose was to test the utility of radio. Originally, four planes were planned. A damaged wing on the NC-2 eliminated it from the flight. The NC-3 almost had a similar fate when one of the wings caught fire in the hangar. The remaining good wing on the NC-2 was substituted and the three seaplanes were ready to take off. Each carried a 500-watt spark transmitter operating on the 425-meter wave, a five-watt radiophone and radio compass, or direction finder.[25] It did not take long for radio to prove its worth to aviation.

Several hours after the take-off from Rockaway, bound up the coast to Newfoundland on the first leg of the flight, one of the three

[25] Radio operators were Harry Sadenwater, NC-1; R. A. Lavender, NC-3; and Herbert C. Rodd, NC-4.

motors of the NC-4 went dead causing a forced landing on the sea off Massachusetts. From the time of take-off, radio compass bearings were taken on the planes by various coastal stations and they showed that the last signals from the NC-4, before the trailing antenna was reeled in, came from off Chatham. When night fell the NC-4 was still lost. At daybreak a destroyer sighted it on the surface of the sea after it had taxied 125 miles in 15 hours to reach Chatham for repairs. Five days later it took off for Trepassey Bay. The NC-1 and NC-3 had hopped off for the Azores.

Before the NC-4 followed, every pound of superfluous weight was removed, many spare parts eliminated, bulkhead doors taken off, heavy tools and an extra fresh water tank were dispensed with. All radio equipment, however, was kept intact for it had proved its value and the greater dangers of the flight were ahead, although more than 25 destroyers and other ships were lined up at regular intervals across the sea to the Azores and on to Portugal.

The NC-1 in a forced landing off the Azores was battered by heavy seas and went to pieces while its crew was rescued by a Greek tramp steamer "Ionia." The same storm damaged the wings of the NC-3 so badly that it could not continue the flight after it taxied into Ponta Delgada. The NC-4, several days behind the other planes, missed the storm and landed safely at Horta in the Azores. After a hop from Horta to Ponta Delgada, there still remained 786 miles to go to complete the Atlantic crossing.

On May 27, the NC-4 in a 9-hour 42-minute flight from the Azores reached Lisbon. Three days later it took to the air again for Plymouth, England, but enroute motor trouble forced a landing on the water and Herbert C. Rodd, the radioman, called for assistance. When the trouble developed the NC-4 headed for the shore near Figueira, Portugal, and came down on the smooth Mondego River. Repairs were quickly made and the plane took off for Ferrol, Spain, for an overnight stop. Early in the morning of May 31 the flight was resumed and arrival at Plymouth was heralded as "achievement of the hitherto impossible in human endeavor!"

Rodd summed up that in the future it would be advisable to have the transmitter powered by other means than a propeller-driven generator, because when the motor went dead the radio did too, except for battery-operated emergency equipment. He found the radio compass invaluable in getting bearings on ships and land stations and in locating harbors. He reported the radio compass signals from ship and land stations were audible for about 50 miles, which was the best distance spanned by such signals during the trip. The range of the five-watt radiophone was about 25 miles at best. In addition to the trailing antenna which was reeled in when landing, there was a so-called skid-fin antenna about 70 feet long, stretched two feet above the top wing.

The NC-4's main transmitter established some new records, one being with Bar Harbor, Maine, which heard the signals up to 800 miles. Rodd described the six-tube amplifier as working perfectly, and "my headphones were buzzing continually" with weather reports, messages, directions for landings, etc.

"At Horta we were entertained at a movie," said Rodd,[26] "and as we entered our box the orchestra played 'The Star Spangled Banner.' And I nearly made a bad blunder. Deafness caused by the constant roar of the motors for 16 hours and signals through a six-step amplifier prevented me from recognizing it and I sat down."

While the NC-boats were making history, Harry G. Hawker, an Englishman, and Lieut. McKenzie Grieve hopped off from Newfoundland in a big Sopwith plane. Shortly after being airborne the undercarriage and wheels were dropped into the sea to conserve weight. No word was heard from them and they were given up for lost until two weeks later when the Danish steamer "Mary," unequipped with radio, reached Scotland with the two airmen, rescued in midocean. Hawker explained that the generator driven by a small propeller was too small and generated insufficient power to spark the wireless.

The next performer was the British dirigible R-34, which crossed the sea in July 1919, from Scotland to Long Island equipped with an arc transmitter. The Zeppelin dirigibles also made dramatic use of radio. Said Leo Freund, a Zeppelin's radioman, "Picture yourself in an airship for three days sailing over the ocean, and the navigator unable to make use of his sextant to determine the position, not being able to see the stars, sun or moon. It would be dangerous to make a trip of this length without radio apparatus."

When the dirigible "Norge" sailed over the top of the world in

[26] *The Wireless Age,* November 1919.

7 3180

LIBRARY
College of St. Francis
JOLIET, ILL.

1926, it flashed the first radio message received directly from over the North Pole. Seventeen years earlier in 1909, Admiral Robert E. Peary had been there, but it took him 153 days to trek back across the ice to the northernmost telegraph station at Indian Harbor, Labrador, from where he announced, "I have the pole, April 6."

Charles A. Lindbergh, on his historic 33.5-hour flight in May 1927, from New York to Paris in the single-engine monoplane, "Spirit of St. Louis," flew without radio.[27] A month later, Admiral Richard E. Byrd's transatlantic 3-engine plane "America" not only lifted a crew of four but also radio equipment. And well that it did, because when lost in the clouds over France and running low in fuel, radio helped to guide it to a forced landing on the beach at Ver Sur Mer. Byrd knew the value of radio for the plane "Josephine Ford," which took him on his first flight to the North Pole in 1926, had carried a 44-meter radio transmitter, although it didn't work too well. But it was different in 1929 when shortwaves from Antarctica announced to the world that Byrd and his pilot Bernt Balchen had flown over the South Pole. Test after test demonstrated how short-waves were aiding the development of aviation-radio. An SOS from a plane, "Dallas Spirit," off Hawaii, was heard in New York 3,500 miles away because the 33-meter wave hopped, skipped and jumped that distance.

Radio continued to adapt itself to aircraft as engineers developed compact, light-weight apparatus operating automatically and at increased efficiency. Radio and the airplane became inseparable. By the sixties, many millions of passengers had flown the Atlantic with radio adding to their comfort and safety.[28]

Great indeed were the advances from the days of the NC-boats to the X-15 rocket plane flying 2,590 miles an hour and up to an altitude over 32 miles with the pilot's voice coming back as if over a telephone. Well that he had the radiophone for he had no time for dots and dashes. Said the pilot, Joseph A. Walker, "When up so high you feel it won't be too long before you're looking down on both

[27] A United States B-58 jet bomber with a three-man crew flew from New York to Paris on May 26, 1961 in 3 hours, 20 minutes averaging about 1,050 miles an hour.

[28] In 1958 aviation first surpassed sea passenger traffic across the Atlantic. In 1968 transatlantic air passengers east- and west-bound totaled 5,752,413, and 376,852 crossed by ship.

sides of the old ball"; he had seen all the California coast, part of Oregon and into Mexico.[29]

Radio had kept pace with aviation. Micro-miniaturization reduced the size of instruments from several cubic feet to the size of a cube of sugar, and even smaller. Modern components became microscopic, some wafer-like and others the size of dice. The transistor and other semi-conductor devices led to what some call "molecular electronics," featuring tiny components and circuits which make electronics ideal for airplanes, missiles, and space vehicles as well as for computers and all phases of communications.

The sixties, therefore, might appropriately be called "The Micro Decade." While microwaves and the maser were leading the advance, so too were micro-devices, so tiny that those who assembled them had to look through powerful microscopes and magnifying glasses to see the specks and gossamer things they were putting together, almost as if to match the invisibility and magic of the electron.

---

[29] The X-15 established an altitude record for winged aircraft in 1963 when it climbed 354,000 feet (67.08 miles) at a speed of 3,866 MPH with pilot Walker at the controls; he achieved a record of 4,104 MPH in 1962. A new record of 4,520 MPH was made in 1966 with Major William H. Knight, U.S.A.F., the pilot.

II   Radio Broadcasting

> Radio is a new agency brought by
> science to our people which may, if
> properly safeguarded, become one of
> our greatest blessings. . . . To the con-
> trol of these channels we should main-
> tain the widest freedom of their use.
>
> —*Calvin Coolidge*

THE STORY OF RADIO BROADCASTING IS ONE OF NATURAL EVOLUTION
of an art and industry born of science. It did not happen overnight,
nor was it any one man's conception.

Broadcasting grew out of wireless. Experimenters were confident
that if dots and dashes could be broadcast so could spoken words and
music. So they set out to develop the wireless telephone, primarily for
point-to-point communication, from shore-to-ship and ship-to-shore.
Their objective was not a medium of mass communication; in fact,
there is no evidence that they thought of such a thing. Their goal was
to make wireless talk. It was no easy job. They had no electron tubes,
no microphones. There were plenty of other missing links.

FESSENDEN POINTED THE WAY

Reginald Aubrey Fessenden might well be called the first American to accept the challenge, although in 1899, as an aide to Edison,
he declared he considered himself "proof against the seduction of
liquid air and wireless telegraphy."

Nevertheless, wireless wooed and won him. He was convinced that wireless had started on the wrong track and he decided to set it right. Tests of what he called "prolonged oscillations," or continuous waves, led him to wireless telephone experiments. On September 28, 1901, he applied for a U.S. patent covering "improvements in apparatus for wireless transmission of electromagnetic waves, said improvements relating more especially to transmission and reproduction of words or audible sounds."

It was explained that a continuous or "undamped" wave is like a note given off by a violin, or a whistle, which can be sustained, and in the case of radio can be molded to carry the voice or music. On the other hand, a discontinuous or highly "damped" wave produced by a spark transmitter comprises a series of intermittent bursts or pulses on the air only when the telegraph key is pressed to form dots and dashes.

At Cobb Island, Maryland, Fessenden set to work and in 1900 succeeded in first transmitting speech by wireless over a distance of a mile, using a 10,000-cycle rotary spark gap. Articulation, however, was not clear. Two years later he obtained better results up to 12 miles by using an arc transmitter, 50,000 cycles per second. In 1903, he increased the range to 25 miles.

The year 1906 was a big one for Fessenden in the realm of wireless telephony. From a new station he erected at Brant Rock, Massachusetts, he "phoned" through the air to Plymouth, Massachusetts, and established conversation between Brant Rock and Jamaica, Long Island, about 200 miles away.

Fessenden was the first to transmit speech across the Atlantic although the feat was not intentional. It happened that conversation between Brant Rock and Plymouth was overheard on several occasions by operators at Machrahanish, Scotland, who recorded the actual words in their log books. And on Christmas Eve 1906, using a high-frequency alternator, Fessenden further demonstrated wireless telephony by broadcasting speech and phonograph music, heard by ships off the Virginia coast; on New Year's Eve reports of reception came in from the West Indies.

Thus Fessenden, known as "the American Marconi" with more than 500 patents to his credit, might well be remembered as the first broadcaster by wireless telephony. He also invented the electrolytic chemical detector and a heterodyne "beat" receiver. But probably his outstanding invention was the high-frequency alternator. After testing two home-built machines he turned the specifications over to the General Electric Company to build a 100,000-cycle alternator; it would have to revolve 20,000 times a second. E. F. W. Alexanderson was assigned the job, and in 1906 a 1-kilowatt 50,000-cycle machine was delivered at Brant Rock.

From Fessenden's day on, the wireless telephone continued to gain attention and captivate the imagination of experimenters, who foresaw the day when passengers aboard ships at sea could talk to their homes and offices.

DEFOREST INVENTED THE AUDION

Ten years had passed since Thomson discovered the electron. In the meantime experimenters were intent upon putting it to work. Gradually the missing electronic links were found. Edison's incandescent lamp had shed light on the subject when, in 1883, he noticed a strange phenomenon inside the glass bulb which became known as "the Edison effect." Electrons streamed as an electrical current from the heated filament and tests showed that they could be caught by a metal plate or cylinder inside the bulb. The filament-plate combination became an object of great scientific interest.

Sir Ambrose Fleming, as electrical advisor of the Edison Electric Light Company of London, was brought into close touch with the problems and mysteries of the incandescent lamp. Exploring "the Edison effect" led him to invent the 2-element (filament and plate) valve, or first electron tube detector for wireless, in 1904.

Two years later, Lee deForest introduced a 3-element (filament, plate and grid) electron tube, which he named the audion. Vastly superior in performance and potential to the 2-element tube, it opened wide the domain of applied electronics and started a revolution in communications. It would detect, amplify and oscillate. No longer would the radiophone be impractical; neither would television or radar, when the world was ready for such services.

Said deForest as he looked at the fragile audion, "Little imagination is required to depict new developments in radiotelephone communications, all of which have lain fallow heretofore, waiting for a simple lamp (electron tube) by which one can speak instead of read."

Invention of the audion was the main step but it created engineering challenges to develop circuits, or hook-ups as they were called, in which the tube could perform; that in itself became a new technology. DeForest made another major contribution when he connected the "B" battery in the circuit with the plate, filament and earphones thereby gaining greatly in amplification. He developed the famous regenerative, or "feedback" circuit, which was used in millions of broadcast receivers.[1]

Sensing the possibilities of radiotelephony, deForest was not content to wait until the audion could be harnessed into a practical radiophone. As a preview, he took an arc transmitter backstage of the Metropolitan Opera House on January 13, 1910, and broadcast opera arias sung by Caruso and other noted artists. Wireless amateurs around the New York area and in Connecticut as well as operators on ships at sea were startled to hear the music.

By 1914, the deForest audion was demonstrated as an oscillator, which made it ideal for voice transmission. It supplanted the arc which had served its purpose in showing that wireless could handle voice and music.

In an empty building on Sedgwick Avenue in the Bronx, at High Bridge, deForest opened a laboratory and factory. And from the antenna tower atop that site many a wireless amateur first heard melodies come through the air and into the earphones.

"The year 1916 was memorable in my life," recalled deForest. "It marked very definitely the actual beginning of planned and systematic radio broadcasting and of the use of the three-electrode tube as transmitter."[2]

He installed a transmitter at the Columbia Gramophone Building on Thirty-eighth Street, New York, which he described as "a very sound business idea, thereby to increase our sale of audions and listening equipment." And in return he agreed to play Columbia recordings; thus broadcasting edged up to sponsorship.

[1] Major Edwin H. Armstrong also claimed the regenerative circuit invention. After years of litigation, the U.S. Supreme Court, in 1934, upheld deForest as the inventor.

[2] *Father of Radio*, autobiography of Lee deForest, (Chicago: Wilcox & Follett Co., 1950), p. 336.

"I still cherished the earlier, quixotic idea," confessed deForest, "that nought but good music and good entertainment or educational matter should go out over the radio."

On election night, November 8, 1916, the High Bridge station broadcast election returns from *The New York American* for six hours. Wireless amateurs were forewarned of the new information service and *The New York Times* reported that "several thousand of them received the news, many through using the newly manufactured wireless telephones."

Incidentally, just before the station signed off for the night at 11 o'clock, the announcer proclaimed the election of Charles Evans Hughes, which the vote from California changed later in the night to Woodrow Wilson.

Up to this time, the main purpose of broadcasting was to stimulate the sale of radiophone apparatus, chiefly to the already established audience comprised of wireless amateurs.

WAR'S IMPACT ON THE RADIOPHONE

World War I put an end to all such activity. Amateur stations were closed and sealed for the duration, but the pace of radiophone development for wartime purposes was accelerated. Hundreds of electron tubes were rigged up for radiotelephony at the U.S. Naval Radio Station, NAA, Arlington, Virginia. Voices broadcast from there in 1915 were picked up in San Francisco and Honolulu, so successfully that a popular songwriter was inspired to compose a new melody, "Hello Hawaii, How Are You?"

RADIO FOR THE HOME

With that chant as an overture, in 1916, David Sarnoff, Assistant Traffic Manager of the Marconi Wireless Telegraph Company of America, sketched the acts for an endless show eventually to open in every city and town and on farms throughout the land. He prepared a memorandum which proposed a "radio music box" that would receive programs broadcast for public information and entertainment. He blueprinted the future that would continually bring new acts into the theatre of the home; everyone would have a front-row seat, and the richest man could not buy for himself what the poorest man got free by radio.

John Philip Sousa would lead his band across the wavelengths. Arturo Toscanini would conduct a great symphony orchestra. Walter Damrosch would direct a Music Appreciation Hour. The Metropolitan Opera would be on the air. A vast audience would listen to Ignace Paderewski at the piano, and hear George Gershwin play his "Rhapsody in Blue." The World Series would be broadcast; so would football and prize fights—there would be no end to the show.

"I have in mind," said Sarnoff, "a plan of development which would make radio a household utility in the same sense as the piano or phonograph. The idea is to bring music into the home by wireless." Then he went on to describe the "music box" and the possibility of broadcasting by radio not only music but lectures and first-hand accounts of events of national importance as they occurred, as well as baseball scores and other information.

When the Radio Corporation of America acquired the American Marconi Company in 1919, Sarnoff became Commercial Manager, and in 1920 he revived the "radio music box idea." He urged considerable experimentation and estimated that with reasonable speed in design and development, a commercial product could be placed on the market within a year or so.

"Should this plan materialize," he said, "it would seem reasonable to expect sales of one million 'radio music boxes' within a period of three years.[3] Roughly estimating the selling price at $75 per set, $75,000,000 can be expected."

When the war ended the radiophone had been greatly improved under the impetus of military exigencies. The time was opportune for a practical test of its appeal to the public.

During the war the Westinghouse Electric & Manufacturing Company by special license from the Government was permitted to build and operate two experimental stations for both telegraphic and telephonic communication. One was located at the company's East Pittsburgh plant and the other in a residential section five miles away, at the home of Frank Conrad, an engineer. When wartime restrictions were removed from amateur radio stations, Conrad took to the air

[3] Actual sales of home radios during the first three years were: 1922, $11 million; 1923, $22.5 million; 1924, $50 million.

with his radiophone transmitter and broadcast phonograph records, talks, and baseball and football scores.

A local department store's newspaper advertisement calling attention to a stock of radio receivers, which could be used to receive the broadcasts from Conrad's station, led to the realization that the efforts being made to develop radiotelephony as a confidential means of communication were wrong, and that instead its field was one of widespread publicity—"instantaneous collective communication."

"It was felt," said H. P. Davis, Vice President of the Westinghouse Company, "that here was something that would make a new public service of a kind certain to create epochal changes in the then accepted everyday affairs, quite as vital as had the introduction of the telephone and telegraph, or the application of electricity to lighting and to power.

"We became convinced that we had in our hands, in this idea, the instrument that would prove to be the greatest and most direct mass communicational and mass educational means that ever appeared. The natural fascination of its mystery coupled with its ability to annihilate distance, would attract interest and open many avenues to bring ease and happiness into human lives. It was obviously a form of service of universal application that could be rendered without favor and without price."[4]

So the decision was made early in 1920 to install a broadcasting station at East Pittsburgh and initiate service as soon as the equipment was ready for operation in the autumn.

Pioneer Broadcasting Stations   On November 2, 1920, KDKA, as the new station was known by its call letters, went on the air with the Harding-Cox election returns which attracted front-page attention in the news. On the following day a daily program service was instituted from 8:30 to 9:30 P.M.

A stampede for wavelengths began; it was like a gold rush in the sky. Convinced by the results at KDKA, Westinghouse put WBZ on the air at Springfield, Massachusetts, in September 1921; WJZ, Newark, New Jersey, on October 12, 1921; and KYW, Chicago, Illinois, on November 11, 1921.

"And where will it end?" asked H. P. Davis in January 1922, "What are the limitations? Who dares to predict? Relays will permit

---

[4] Lecture at Harvard Graduate School of Business, April 21, 1928.

one station to pass its message on to another, and we may easily expect to hear at an outlying farm in Maine some great artist singing into a microphone many thousand miles away. A receiving set in every home—in every hotel room—in every school—in every hospital room. Why not? It is not so much a question of possibility—it is rather a question of *how soon.*"

While KDKA claimed the honor of being "the world's first broadcasting station," others claimed priority. *The Detroit News* declared that it "opened the original radio broadcasting station" on August 20, 1920; the call was WWJ.

Said Lee deForest from whose company *The Detroit News* purchased the installation:[5]

". . . To *The Detroit News* belongs the honor of having first foreseen the possibilities of radio broadcasting as a means for widespread dissemination of news and its value as an adjunct to the newspaper. As early as 1919, when a representative of the deForest Radio Company brought first to the attention of the management of *The Detroit News* the possibilities of broadcasting news from their offices, your paper was prompt to grasp the idea and quickly decided to try the experiment. Success of your truly pioneer installation was instantaneous."

A look at government records to find the answer regarding the oldest broadcasting station reveals that the first regular broadcast license was issued to WBZ, Springfield, Massachusetts, on September 15, 1921. WWJ, Detroit, obtained its first regular license on October 13, 1921. KDKA, Pittsburgh, however, held an experimental license prior to that time which permitted it to conduct "radiotelephone" tests, and that license was dated October 27, 1920; KDKA's license for regular broadcasting was issued on November 7, 1921. The answer to the question of who was the first broadcaster depends upon whether one wants to make the selection based upon experimental radiophone broadcasts, or the date of the station's first license for regular broadcasting.

In any event, 1920 was the kick-off year. Overnight it seemed everybody wanted to listen in. Radio sets were few and far between; only the wireless amateurs had the means of listening.

The Radio "Craze" Spread While the electrical manufacturers were "tooling up," thousands of young and old "Marconis" began to

[5] Message sent by Lee deForest to *The Detroit News* on the 12th anniversary of WWJ, August 20, 1932.

build their own "radio music boxes"—crystal detector sets. They were relatively simple and the components were easy to make, except the earphones. Wire wound around a cardboard tube, or wrapped around a cylindrical cereal box, sufficed for a tuning coil; tinfoil from the florist facilitated for a condenser. The crystal detector was simple—a piece of galena, silicon or carborundum, held in a metal cup or clip, so that a tiny wire could be moved across the surface of the mineral until the most sensitive spot was found for good reception. Thus the popular name "catwhisker" detector was coined.

Finally, in 1921–1922, "radio music boxes" appeared on the market featuring electron tubes, some with one tube and others with two or three for amplification. Up to this time, RCA operations related chiefly to radio communication, but it now entered the sales and merchandising field to market the radio sets produced by its manufacturing associates; Westinghouse named its set "Aeriola," and General Electric introduced a "Radiola for every purpose."[6]

Every day seemed to witness some new "first" in broadcasting— dance bands, church services, lectures, championship fights and anything that qualified through sound for the public interest. It was not long, however, before entertainers began to frown upon this new-fangled monster. If they could be heard in the home via radio why would anyone pay to attend their concerts and performances?

As Paul Althouse, star of the Metropolitan Opera, put it, "It's bad enough to sing to a big audience but—ye gods, to a piece of tin!"[7]

The New York broadcasters soon learned that artists were reluctant to go to microphones across the Hudson or to out-of-town studios. It took a lot of coaxing to get Charlie Chaplin, Bebe Daniels, and others to take time to limousine and ferry across the Hudson. It was not long before WJZ shifted its studios from Newark to Aeolian Hall on Forty-Second Street, and WOR moved to Broadway, making it more convenient for talent to appear. Several stations also moved their transmitters to Manhattan Island, but later moved back to the

[6] As a means of acquiring manufacturing facilities, as well as an established phonograph and record business, RCA, in 1929, acquired the Victor Talking Machine Co. And in 1932, RCA became a completely self-contained organization, entirely independent of the companies with which it was associated in early radio development.

[7] *You're on the Air,* by Graham McNamee (New York: Harper & Brothers, 1926), p. 107.

hinterlands to gain transmission advantages that provided better coverage free from absorption of the waves by New York's skyscrapers.

Battle of the Century   Up to this time, no event except the Harding-Cox election returns touched off the radio craze as did the blow-by-blow broadcast of the Dempsey-Carpentier heavy-weight fight on July 2, 1921, at Boyle's Thirty Acres, Jersey City. When the gong clanged the beginning of the first round it signaled a new era in radio.

Wherever there was a radio set in the Metropolitan area it was turned to the 1,600 meter wave of station WJY set up for the occasion in the D.L.&W. Railroad terminal in Hoboken, 2.5 miles away from the battle scene. A radiophone transmitter, borrowed from the Navy, was connected by telephone wire to the telephone mouthpiece at the ringside. It was estimated that the audience ranged from 200 thousand to 300 thousand listening in at homes, theatres, lodge halls, ballrooms, parks and barns from Maine to Florida and as far inland as West Virginia.

Most of the receiving sets were operated by wireless amateurs alerted in advance that the "Battle of the Century" was to be on the air. Typical of the installations was one at Wein's department store in Port Chester, New York, where a big morning-glory shaped phonograph horn protruded from a second-floor window. The street was massed with such a crowd that no traffic, not even a street car, could get through until the radio bellowed, "—nine, ten! The fight is over! Jack Dempsey remains heavyweight champion of the world!"

Major J. Andrew White, the ringside announcer, was himself broadcast into fame that afternoon; he became radio's foremost sports announcer for many years. The big wooden saucer at the Acres held about 91 thousand fight fans, and when many of them returned to New York from the hot, dusty battle ground they were astounded when those who had heard the fight on the radio knew more about it than they did.

HISTORIC RADIOTELEPHONE TESTS

Radio and electrical companies quickly stepped up their activities in various phases of broadcasting. The American Telephone & Telegraph Company, along with the Bell Laboratories, always interested in whatever related to voice communications, lost no time in exploring the various facets of radiotelephony—present and future.

As early as 1919, an expenditure of $500,000 had been authorized by A.T.&T. for a wireless development including a $360,000-marine (radiophone) transmitter at Deal Beach, New Jersey. As a result, in 1922 the first ship-to-shore two-way radio conversation was conducted between the Bell System's station at Deal Beach and the S. S. "America," 40 miles at sea. Also, the S. S. "Gloucester" off the New Jersey coast talked to Deal Beach which relayed the voices via wire to Long Beach, California, and from there by radiophone to the Catalina Islands.

Then to explore the field of broadcasting, A.T.&T., in July 1922, put a 500-watt station (WBAY) on the air in New York. A month later its activities were transferred to WEAF, atop the Western Electric building on West Street.

By this time there was no doubt that radio was in the home to stay. Fan letters from all parts of the country told the story. One to WEAF said,[8] "It is 5:25 P.M.,—you have just finished broadcasting; you have practically finished breaking up a happy home. Our set was installed last evening. Today, my wife has not left her chair, listening all day. Our apartment has not been cleaned, the beds are not made, the baby not bathed—and no dinner ready for me."

WHO WOULD PAY FOR BROADCASTING?

More and more, practical business men were asking, "Who is to pay for broadcasting?"

Any number of plans were offered but none seemed sufficiently comprehensive or capable of withstanding the test of real analysis; in fact, most of the proposals called for voluntary payment by the public.

A system of "narrowcasting" was suggested whereby a coin box attached to the radio set would collect revenue from the listener. "Wired-wireless" was also offered as a means of sending the programs over telephone wires to receiving sets in the home on a rental

[8] *Commercial Broadcasting Pioneer: the* WEAF *Experiment 1922–1926,* (Cambridge: Harvard University Press, 1946), p. 90.

basis. But advocates of free radio would have nothing to do with such ideas—broadcasts must be as free to the public as the air; they declared that radio's big advantage was in its universality and in its ability to reach everybody, everywhere and anywhere.

Another proposal called for a chain of superpower stations across the nation over which new talent as well as noted artists would consider it "a mark of distinction" to perform gratis for prestige and publicity; they would gain an audience in one evening larger than they could hope to have in a lifetime on the theatre or concert stage. But that idea didn't take; as one artist put it, "prestige doesn't buy the baby shoes."

As the months went on, the broadcasters were warned that they had better stop pipe-dreaming and solve the problem, otherwise the whole industry would find itself founded on sand and collapse for lack of support.

If entertainers refused to go on the air without payment under "the prestige plan," then it was suggested, in 1923, that the radio manufacturers and distributors contribute from $2 million to $5 million a year to pay the talent. Surely, it was pointed out, a half-billion dollar industry could afford to tax itself 1 or 2 per cent.

Finally, the question was settled once and for all at station WEAF, New York. This is how the riddle was solved: The Western Electric Company, manufacturing subsidiary of A.T.&T., had requests for more than 200 broadcast transmitters. Foreseeing the high costs of broadcasting and increased competition, aside from the necessity to pay talent together with limited program material in some communities, the A.T.&T. said in effect to those who wanted to own and operate stations: "Why not share some of our broadcast time? Why not buy a time period from us and sponsor a program of entertainment over our station WEAF?"

The First Sponsors The invitation caught on. As a result, the first commercial broadcast is logged in the annals of broadcasting as of August 28, 1922, when the Queensboro Corporation sponsored a ten-minute talk on Housing by M. H. Blackwell; the cost $100. He told about tenant-owned apartments at Jackson Heights, and after several such broadcasts, sales amounting to several thousand dollars were reported.

Apparently mindful of Benjamin Franklin's admonition, "If you are not getting into the home with what you have to *say,* you will never get in with what you have to *sell,"* the next sponsors were the Tidewater Oil Company and the American Express Company. Both made experimental broadcast announcements on September 21.

"Should Radio Be Used for Advertising?" was the subject of a timely article in *Radio Broadcast* magazine, in which the author, Joseph H. Jackson, prophetically said:[9]

"Driblets of advertising, most of it indirect so far, to be sure, but still unmistakable, are floating through the ether every day. Concerts are seasoned here and there with a dash of advertising paprika. . . . More of this thing may be expected. And once the avalanche gets a good start, nothing short of an Act of Congress will suffice to stop it."

Commercial sponsorship was on the way whether the pure in heart, who didn't want broadcasting "tainted by advertising," liked it or not. There were some who faithfully clung to the ideal of "cultural" radio; they held to the conviction that advertising was "a commercial snake with rattlers in the garden of entertainment."

Be that as it may, the "gold rush" was on. Broadcasting stations had increased from about three to 595 between January 1922 to January 1923. Hundreds of others were on the way. Radio, it was declared, had passed from a toy to a national joy. Broadcasters were no longer in radio for fun, novelty and romance. They had to make money to survive; their philanthropic days were over. Advertising was their salvation and they welcomed with open arms the commercial sponsors who unlocked the economic door.

There came another revelation. Marion Davies, movie star, sponsored by Mineralava Face Clay, delivered a series of ten-minute talks over WEAF, and offered an autographed photograph. The first week brought 6,783 requests. Surely that was proof enough to convince advertisers that an invisible audience was really at hand.

Early in 1923, "The Silver Masked Tenor" and the "Silvertown Orchestra" were on the air, sponsored by the B. F. Goodrich Rubber Company, advertising Silvertown tires. Results were so good from a sales standpoint that Goodrich gave up the idea of buying its own station.

---

[9] November, 1922.

It was about this time that Graham McNamee, a concert singer, happened to stroll down Dey Street and went into WEAF to see what a radio studio looked like. The manager asked him how he would like to try the work. Since the hours were short and wouldn't interfere with his music he decided to try radio; he was allowed to do some simple announcing, and before long his voice enlivened radio as never before and made it seem real. As Heywood Broun remarked, "McNamee generated in himself the same excitement which the game churned up in the crowd. . . . He gave out vividly a sense of movement and of feeling. Of such is the kingdom of art."

Extending their activities into public affairs, the broadcasters put the Republican and Democratic Conventions on the air in 1924 with McNamee at the microphone. The New England twang of Calvin Coolidge was heard afar, although he was not the first President to broadcast. Harding was the first while in office in 1921; Wilson in 1923 after retiring from the White House. Hoover took to the air frequently while Alfred E. Smith enlivened the campaign of 1928 by his use of what he called "the rad-dio"; he referred to the microphone as "the pie-plate." Roosevelt's fireside chats became historic, and from then on radio has been at the service of the President.

After the first five years of broadcasting, home-set building tapered off, largely because the electron tube circuits were more complex, especially the Armstrong superheterodyne, which came on the market in March 1924, and the Hazeltine neutrodyne. There were plenty of "radio music boxes" on sale at reasonable prices. Crystal detector sets were on the wane by 1924. By the time the all-electric set was introduced (1926) the era of home-set building was over and the earphones had given way to the loudspeaker in various forms and shapes from the gooseneck horn to paper cones and fancy boxes.

Setting a Precedent   Such were the radios that tuned in the Dempsey-Tunney fight on September 23, 1926. At the Philadelphia ringside, Major J. Andrew White was the blow-by-blow announcer over a hook-up of thirty stations—a record for that day. Several precedents were set in sports broadcasting, including sponsorship by the Royal Typewriter Company, which was commended for never mentioning the trade name or product during the 45-minute battle. A critic observed that the commercials were worked in a fair number of

times before and after the fight, but he added, "Probably not too
often when you consider it was costing about $1,000 per mention";
he called it "a commmendable exhibition of restraint."[10]

He deplored, however, the precedent established in purchasing the
broadcasting rights for a sports event at a cost reported to be
$35,000, including the expense of broadcasting. It was fraught with
dangerous consequences,[11] because, the critic said, "the payment of
large fees for broadcasting sports, which themselves profit by broadcasting, is unfair and unwarranted; the prosperity of boxing has been
tremendously helped by the impetus which radio has given it. . . . It
is our guess that the Dempsey-Tunney fight had the largest audience
ever attracted by a single broadcast, which means, the largest audience in the history of the world."

ARTISTS' SHYNESS VANISHES

More than ever the broadcasters craved big names and big events
to gain more listeners. Publicity was no longer bait; neither was the
listing of performers' names in the daily radio programs of the newspapers. If a sponsor would pay $35,000 for a prize fight, why
shouldn't a singer or comedian receive at least $1,000? So perhaps to
bring it about, they should shy from the microphone until commercial
sponsors crossed their palms with silver.

Shyness vanished on New Year's 1925 when the "Atwater Kent
Hour," adopting the "silver method," enticed John McCormack,
tenor, and Lucrezia Bori, soprano, to sing for the microphone. That
evening marked the first broadcast sanctioned for artists of the Victor
Talking Machine Company, and it was also considered as a nod of
approval by the Metropolitan Opera. Other artists no longer hesitated
for they reasoned what was good enough for McCormack and Bori
was good enough for them. Jascha Heifetz with his famous violin was

[10] *Radio Broadcast*, December, 1926.

[11] Closed-circuit TV theatre and arena rights to the Patterson-Johansson
championship fight, March 13, 1961, at Miami Beach, Florida, were reported
at $1,800,000. Radio rights were $300,000 and community TV antenna system
rights were $60,000. The fight was not telecast but the pictures were "piped"
into 207 theatres and arenas throughout the country which provided 756,195
seats. TV pictures were also handled by thirty community antenna systems serving about 100,000 TV sets.

paid $15,000 for his half-hour radio debut one Sunday night, advertising a brand of coffee.

All performers found themselves in a new and strange world—in the heavily padded and draped studios there was no noticeable reverberation. McCormack at rehearsal exclaimed, "This is dead! I can never sing here!" And Bori added, "Why, I can't even hear myself or tell what I am doing. Does it sound that way outside?"

They missed the warmth and encouragement of an audience. Will Rogers, facing the dead wall and "mike" seemed lost; he kept turning around to look at the monitor booth for some approval—for an inkling that he was being heard; he needed audience response, laughter and applause. Sensing the situation Graham McNamee left the control booth and went into the studio to provide the response an actor needs.

RADIO'S ILLUSION

Radio thrived on illusion. The roots of its popularity were implanted in the listener's imagination. Showmen soon learned that their trick for success was to play on the imagination with sound, to create a mental picture so that every listener would revel in a pleasant illusion. By words and sound effects they cleverly created a fanciful theatre. Every listener was part of it in his own mind; they all "lived" with Amos 'n' Andy as they visualized them and each one probably "saw" them differently as he imagined the scene.

On one occasion it was announced that the radio "Showboat" would put in at Erie, Pennsylvania, the next week. Several hundred people went to the waterfront that night to see the showboat which they had followed on the radio week after week. But, of course, it came in only on the radio; those at home "saw" it, while those at the dock missed it. There was no actual showboat; only actors hovered around a microphone reading scripts abetted by sound effects that made it seem real in the radio land of make-believe.

The most dramatic broadcast that revealed the power of radio coupled with illusion was based on *War of the Worlds* by H. G. Wells, featuring Orson Welles.[12] It caused mass hysteria among

[12] October 30, 1938.

radio listeners who were led to believe "a gas raid from Mars" was actually striking America. Listeners heard screams and shrieks followed by a terrific explosion and then the excited voice of the announcer breathlessly exclaiming: "Now the whole field's caught on fire. The woods, the barns, the gas tanks of the automobiles—it's spreading everywhere. It's coming this way! About twenty yards to my right!" Then the microphone was heard to crash—and silence!

Telephone switchboards of police, newspapers and broadcasters were jammed. Households were disrupted; adults and children were frightened by what promised to be remembered as the most horrifying bedtime story ever broadcast in the United States. The playwright, abetted by the dramatic voice of Orson Welles, had fired imaginations of radio listeners from coast to coast. It was unbelievable, yet thousands were led to believe that Martians were attacking the earth.

Broadcasters learned once and for all that melodrama dressed up as a current event was dangerous; that fiction, fables and fantasy should not be dramatized to simulate news.

RADIO'S EFFECT ON NEWSPAPERS

The newspapers began to ponder their role in creating the new stentorian giant that was even casting news—their product—to the winds. Who would want to read the newspapers? And didn't this new medium, already heralded as a new dimension in journalism, threaten to divert revenue from newspaper advertising? Was the press helping to create a Frankenstein?

"Newspaper publishers are making a wholly gratuitous contribution to a competing medium," warned *Editor & Publisher,* "by permitting it to broadcast news and headlines."

Said Karl Bickel, president of the United Press: "As far as news is concerned, radio is a great bulletin board and by its limitations it can never be more than that."[13]

"Broadcasting is not a rival of the printed word," said Merlin H. Aylesworth, first president of the National Broadcasting Company. "It is far more elementary in conception than the printed word; it is a vehicle for spontaneous thought and for expression of personality. It is born, it lives and dies in the space of a second or less."[14]

[13] *Radio in Advertising.* (New York: Harper & Brothers, 1931), p. 30.
[14] *Ibid.,* p. 33.

Many publishers looked upon radio as a menace yet they helped popularize it by printing daily programs gratis as well as featuring news about broadcasting. Time and time again the publishers decided to eliminate listing the radio programs, or make the broadcasters pay for the service. The public, however, complained bitterly and finally one paper would restore the listings, forcing the others to reinstate them.

Some solace was found in a prediction by H. G. Wells that radio would pass into oblivion quicker than mahjong and the crossword puzzle. Nevertheless, cries continued that radio threatened the newspapers and magazines and even opera and the movies. To gain more from sponsorship advertisers linked commercial names to programs, dance bands, etc. For example, there was the "Pepsodent Hour," "The Gold Dust Twins," "Texaco Firechief," "Maxwell House Showboat," "Lucky Strike Orchestra," "Camel Caravan," "Cliquot Club Eskimos," "Ipana Troubadours," etc. Finally, newspaper publishers were so incensed that they ruled out all such listings; they would tolerate only "Concert Orchestra," or the name of the conductor such as "Toscanini Orchestra"; or the name of the performer, such as Ed Wynn in place of "Texaco Firechief."

Carr V. Van Anda, Managing Editor of *The New York Times,* who instituted radio program listings in that paper in 1921, and introduced radio news pages in the Sunday *Times* in May 1922, declared, "The radio program listings are more important than the stock market listings, because more people are interested in radio."

Van Anda was a staunch advocate of radio and his astute news sense enabled him to see where radio was headed. In 1923, he proposed that *The Times* erect the most powerful broadcasting station in the country to feature news. He discussed the idea with Alfred N. Goldsmith, then Professor of Electrical Engineering at the College of the City of New York, who said the super-power proposal was practical but, in order to minimize interference, he recommended that the transmitter be located in New Jersey with the studio in New York. Adolph S. Ochs, the publisher, adhered to the policy that *The Times* was in the publishing business, not broadcasting. Twenty years later, in 1944, *The Times* bought station WQXR for a price reported as $1 million.

Publishers began to recognize the value of radio as a supplement to publishing, and they acquired broadcasting stations and television too.[15] Both had distinct advantages of offering services the linotypes and presses were not capable of supplying.

Radio can broadcast news the instant it happens from around the world. It cannot be as detailed or complete as a newspaper; its meteoric news flashes vanish into space the instant they are uttered. With the development of tape recording, radio programs could be preserved for the future. Whether taped or not, the broadcast must be heard at the instant and the listener has no permanent record as provided by a newspaper. Print, less demanding from the standpoint of time, can be read at will. It is a permanent record for posterity.

Publishers found some relief in the fact that radio was like a shooting star, seen only by those whose eyes happened to be turned in its direction. Similarly, the fleeting broadcast bulletin is heard only by ears tuned to its path through space. The radio news flash is a flying particle, perhaps a tantalizing morsel that is likely to send listeners for a newspaper to get the full story. Indeed, broadcasting often whets the appetite for news as it dramatizes local, national and international affairs. News on the air advertises news.

Listening demands greater concentration than reading; radio calls for immediate attention. The newspaper can be laid aside and be picked up again after the telephone or doorbell is answered, or after the baby stops crying. Broadcasts paced by the clock must be tuned in on the spur of the moment. Like time and tide, broadcasts do not wait.

Broadcasts enhance the prestige of newspapers. The obligation of the newspaper to be judicial becomes greater than ever—radio listeners as well as TV viewers often read to confirm what they have heard and seen, in much the same way that a baseball fan buys a newspaper to read about the game he attended.

Brevity of the weather report makes it ideal for broadcasting. The bulletin, or news flash, however, is not broadcasting's only link with the press. News frequently is born at the microphone and TV camera. President Roosevelt's fireside chats are a good example, as are presidential news conferences. The farewell address of King Edward VIII is written in the records as one of the most dramatic news broadcasts of all time—the epitome of all that radio can be.

[15] 391 newspapers and 40 national magazines were associated in ownership of radio and TV stations, according to the 1969 Broadcasting Yearbook.

There seems no doubt that broadcasting has made the public more news conscious. Marconi was asked if he thought broadcasting would supplant newspapers.

"No," he replied. "My wife hears something on the radio and wonders if I heard it, because she doesn't remember it all, it went so fast. So, if I am interested, I go to the newsstand for a newspaper. Naturally, if neither of us happened to be listening when the news was on the air, we missed it, so we buy a paper, and it has pictures too. We can read it when we find the leisure."[16]

From the publisher's standpoint, radio diverted advertising revenue from the press; at first it might be 25 per cent, but gradually it would settle down to perhaps 10 per cent. But while such a reduction in revenue was taking place, broadcasting, as a new medium of advertising, stimulated public appeal and increased sales, and it broadened markets. That meant greater advertising expenditures, not only for broadcasting, but for newspapers and other media.

Fully aware of the value of the press and other sources of amusement, entertainment and information, leaders in the field of broadcasting continually declared that the electronic art was no threat to them, except to those decadent enough to deserve oblivion. Radio, they asserted, was a stimulant for what was new and better; it would educate people in that direction and stimulate interest and desire for the better things of life. Producers and showmen were told that insofar as entertainment catered to those desires and instincts, they had nothing to fear and everything to gain from radio. But they would have to accept the new and energetically reconstruct the old, a combination which would lead to progress.

Even radio broadcasting in its formative years was continually changing to enhance itself as a distinctive and encompassing medium, wholly personal in appeal. While many new stations took to the air, some of the pioneers changed hands.

ADVENT OF THE NETWORK

Station WEAF, at which the answer to the million-dollar question was found and proved to be correct, was purchased for $1 million in 1926 by the Radio Corporation of America from A.T.&T., which

16 Interviewed by the author, September 1933.

decided to retire from the broadcasting business. At the same time RCA announced formation of the National Broadcasting Company with WEAF as a key station. It was estimated that five million homes had radios and 21 million homes remained to be equipped, for radio was "no longer a plaything but an instrument of service." The purpose of NBC would be "to provide the best programs available for broadcasting in the United States." That opened the era of national network broadcasting.

SHORTWAVES TO THE FORE

Shortwaves became the key to international radio. They are to radio what good roads and super-highways are to the automobile—they can get places fast. Shortwaves made it as easy to communicate across the hemispheres as talking across a city in the early twenties.

As sort of a final fling for the longer waves, a dance band in London played over the 1,600-meter channel for reception at Houlton, Maine, from where the melodies were relayed over wire to New York for rebroadcast by WJZ. It was December 29, and a grand climax for broadcasting in 1923. At that time, Donald B. MacMillan, exploring in the Arctic used shortwaves from his ship, the "Bowdoin," to communicate with New York, Chicago, and other cities.

Shortwaves seemed to come from everywhere in 1924; London, Calcutta and Cape Town picked up broadcasts from KDKA in Pittsburgh. Marconi talked to Australia from his yacht off England, and from the Mediterranean he conversed with Syria over the 32-meter wave in broad daylight.

When a transatlantic radiotelephone circuit was opened in January 1927, Geoffrey Dawson, editor of the *Times* of London talked with Adolph S. Ochs, publisher of *The New York Times,* who declared, "Who now has the temerity to say that prayers are not heard in Heaven?"

The thirties might well be called "the shortwave decade" as shortwaves made radio more and more international. Expansion brought obsolescence of long-wave equipment and lofty towers; it also brought surprises as the march of progress extended to ultra-shortwaves.

FM INTRODUCED AS STATICLESS

Major Edwin H. Armstrong, a disciple of ultra-shortwaves, tossed a bombshell into the firmament of broadcasting in 1935 when he demonstrated a new system—frequency modulation, popularly called FM—at a meeting of the Institute of Radio Engineers. Superb in tonal excellence, and out of range of atmospheric disturbances as well as other extraneous noises, it was staticless. He urged broadcasters to take to FM.

They argued, however, that such a revolutionary change-over at that date would be like changing the width of all railroad tracks throughout the country to accommodate a new locomotive. Because amplitude, or AM broadcasting was so well established, it would not be economical to shift. On the other hand, they might supplement the AM stations with FM transmitters, especially if enough FM receivers were in the field to warrant such a change.

Meaning of AM and FM   When a broadcast transmitter is in operation, it transmits energy continuously whether or not a signal is being sent. This continuous transmission is called the carrier wave; it carries the signal. When a signal is put on the air, the pattern of the carrier wave is altered to be compatible with the pattern of the signal, whatever sound is impinged upon the microphone. This variation in the pattern of the carrier wave is called *modulation.*

Several ways can be used to modulate the carrier wave; in broadcasting, two methods—amplitude modulation (AM) and frequency modulation (FM)—are used.

Amplitude in a water wave is its height or depth compared to the normal water level. Similarly, amplitude in a radio carrier wave is its height or depth compared to its normal level. In *amplitude modulation,* the carrier wave is maintained at a constant frequency while its amplitude is varied.

Frequency in water waves is the measurement of the number of waves passing one point in a given time. Similarly, frequency in radio waves is the measurement of the number of radio waves or cycles passing a point in a given time. In *frequency modulation,* the carrier wave is maintained at a constant amplitude and its frequency is varied.

In both methods, it is the job of the microphone to translate sound into electrical energy, identical with the original sound pattern. The electrical impulses from the microphone are highly amplified and fed into the transmitter, the unmodulated carrier wave of which is on the air awaiting the signal to be imposed on it. Thus the wave is modulated and the pattern of sound energy is transferred into electrical energy and into controlled electromagnetic waves.

Every radio station is assigned a frequency or channel, popularly known in the earlier days of radio as a wavelength. Standard AM broadcasting stations are allocated the frequency band 540 to 1,600 kilocycles, and since each channel occupies 10 kilocycles there is space for 107 stations.

Broadcasting first developed on the basis of amplitude modulation, but radio engineers were at least theoretically aware that similar results could be achieved by frequency modulation which presented more technological problems. While sound reproduction in AM is limited to frequencies up to 5,000 cycles because of channel capacity, in FM the channel width of 200 kilocycles permits reproduction up to 15,000 cycles or over, providing a much wider range of sound. Furthermore, since FM is not dependent upon signal amplitude but on signal frequency, static and other extraneous noises cannot readily intrude to contaminate the broadcast—thus higher fidelity.

To prove the point, Armstrong built a 40-kilowatt FM station at Alpine, New Jersey, the 400-foot tower of which was destined to become his monument.[17] He put Alpine on the air in January 1939, on the 7.5-meter wave, and invited members of the press to his apartment in River House on Manhattan Island, to hear it; he said he was sure it was the basis of a real news story.

"I don't expect any one to believe it until he hears it," said Armstrong. "You've just got to hear it to believe such true tone and clarity possible."

The next day, *The New York Times* reported:[18] "Not only has the Major and his engineering crew succeeded in dodging static but they gave to radio a remarkable system of pure-toned broadcasting. So realistic is the music that it is as if the listener were sitting amid

[17] Died, February 1, 1954.
[18] January 22, 1939.

the musicians instead of miles away. The wave is surprisingly silent except for the program."

Armstrong in his living room picked up the telephone and called Alpine to begin the demonstration. Near the microphone water was poured into a glass tumbler; the tinkle was as lifelike as if the water were being poured in the room where the receiver was located. To further show the uncanny realism of FM he asked the operator to strike a paper match in front of the microphone and then a wooden match. It was amazing how this new form of radio showed the difference; scarcely a sound except the scratch when the paper match was lit, but as the wooden stick was struck, the burning of the splinter sounded as if close to the ear.

"It is simple when you know how to do it," explained the Major. "People who learned about radio before broadcasting came along know that new things are possible; they believe in the system (FM). But most of the newcomers in the field since 1920 see radio as a cut-and-dried affair, the investment in which cannot be affected by an improved structure. But we oldtimers remember how the spark gaps went out, to be replaced by the vacuum tube, etc.

"Of course, the present broadcasting system is not to be destroyed. Eventually, however, as I foresee it, listeners will have a receiver capable of selecting the same program on either the standard broadcast waves, or on ultra-shortwaves. Because of the wonderful tone on the ultra-short, the listeners no doubt will favor them."[19]

Armstrong, like Marconi and the majority of inventors, was beset by claims and counter-claims of priority. Had not Cornelius D. Ehret, in 1902, applied for a patent on frequency modulation, which was granted in 1905? And there were others who claimed to be first in FM.

Be that as it may, Armstrong did more than any other individual to put FM in the news, developed it and crusaded for it. He was a victim of time. FM was too late to become the primary national system of broadcasting; too many AM stations were on the air and millions of radio sets in operation were designed for AM, not FM. Had FM been ready for general public service before 1923, it might have overcome

[19] Total radio sets in the United States in 1969 were estimated at 303,400,000 including about 75 million auto radios. Industry consensus put the FM set figure at about 50 million.

the handicap—but it was too late in 1939 for a quick change-over of everything. True to Armstrong's prediction, however, FM stations were established alongside the older AM system, and FM won its rightful place in broadcasting as well as in other fields of radio communication. It was adopted as the sound portion of television as well as for use in radar, telemetry, and other non-broadcast services. And when the FCC approved stereophonic "multiplex" broadcasting with its added "live" performance standards, effective on June 1, 1961, another boon for FM was foreseen.

Stereo FM This system features two signals, varying slightly in frequency, both carried on a single FM wave that transmits the sound from two or more microphones placed at different locations in front of an orchestra. The stereophonic FM receiver is equipped with two loudspeakers each of which handles one of the stereo signals. Existing FM radios, of which there are more than 50 million, will reproduce the stereo programs, but the two signals are blended monophonically through a single loudspeaker unless an adaptor is used. Stereophonic records and tapes can also be broadcast.

Stereo in effect is space identification, or localization of the sound sources; it reveals where the instruments are located—the violins in one section of the orchestra, the woodwinds in another, etc. It creates the impression of the large orchestral area from which the sound comes rather than a tunnel effect as from a single loudspeaker. Enthusiastic about this new trend in broadcasting, one manufacturer evaluates it as the greatest thing for the electronic industry in the realm of radio entertainment since TV. FM's share of the consumer market registers new highs from year to year. The majority of radio-TV combinations and radio-phonographs contain FM-AM tuners. Portable transistor radios are recognized as far and away the most popular type followed by auto, clock, and table models that feature FM.

Time was upholding Armstrong's faith in FM, although in the beginning the factor of timeliness that contributed so much to Marconi's success did not favor him. Nevertheless, his invention of the super-heterodyne receiver universally used in commercial radios, home radios and television sets, alone would have won him a prominent place in radio's Hall of Fame. His patent application for "the super," filed in 1919, was granted in 1920, and when placed on the market in 1924 it was heralded as a masterpiece and "the peer of all radios."

TRANSISTORS INVENTED

Every year throughout the history of radio brought some new development—some fit easily into the existing systems; others more revolutionary came at less frequent intervals. For example, the transistor, a semiconductor non-electron tube device, was developed by scientists of the Bell Laboratories in 1948.[20] It was the key to a new branch of electronic science—solid-state electronics, so called because electrons could be liberated from tiny chips of solid materials such as germanium, silicon, and other metals. No longer was the electron tube with its heated cathode exclusively essential as a device to free electrons from a placid state and harness them for useful purposes. Now a semiconductor performed the trick with marked simplicity and without generating heat. Rugged and less fragile than a glass bulb, tiny in size, and needing little current for operation, the semiconductors opened the way for radically new compact and portable designs of radios, television sets, and other electronic instruments and communication systems.

The semiconductors comprise materials that are intermediate in their characteristics between conductors (metals) and insulators (porcelain or glass). They conduct more than insulators, but less than metals, and that is why they are called semiconductors. They are made of carefully crystallized materials, such as germanium or silicon, to which is added a very small precisely selected amount of impurities, such as arsenic. Different types of semiconductors can be combined to form diodes (rectifiers) or even triodes, functioning like the 3-element electron tube.

These mighty mites of electronics, some the size of a kernel of corn or even smaller, also had timeliness on their side; they came upon the scene when electronic computers, missiles, satellites, and spaceships needed the advantages they offer. And since the semiconductor is a detector, amplifier, and oscillator, for the first time the electron tube

[20] Dr. William Shockley, Dr. John Bardeen, and Dr. Walter H. Brattin.

faced competition. Now radios could be made compact, light-weight, and pocket-size, no larger than a pack of cigarettes. Indeed, the transistor and its offsprings such as the tunnel diode had a multiplicity of uses throughout electronic communications.[21]

Lee deForest was asked in 1958 what he thought about the turn of events; what would have happened had the transistor appeared in the early years of the century?

"Well," he said with a reflective smile, "I might never have invented the audion."

Be that as it may, time proved beyond all doubt that it was a good thing that he did invent the audion for it revolutionized electronic communications—radio as well as telephone; it became the heart of radio broadcasting as well as television and radar. Time will show that both electron tubes and semiconductors have their places in communications as well as industrial electronics. Each will enact roles the other cannot perform to the same advantage. For the Space Age, of course, the semiconductors are ideal for vehicles that carry electronic instruments into the ionosphere under most rigorous conditions. And they are tiny, almost weightless, consume little current, withstand shock and vibration, and are long-lived despite the tremendous temperature changes and radiation effects encountered at satellitic altitudes.

Again, as Armstrong had noted, oldtimers recalled how the spark gaps and crystal detectors went into discard when the electron tube appeared. And the irony now was that the crystal—a tiny speck of germanium or silicon—was back in a new form, threatening revenge for what the tube had done to the crystals that detected wireless signals and broadcasts by the pioneers.

Scientifically it would always be the same—something new and more efficient, designed to extend electronics into new and fertile fields of service.

[21] In 21 years the transistor and its solid-state relatives became a $10-billion industry and helped lift the electronics industry from fortieth in annual dollar volume to fourth with $30 billion in sales.

III Radiophoto

> One picture is worth ten thousand words.
> —*Chinese proverb*

RADIO'S SUCCESS IN BROADCASTING VOICES AND MUSIC LED TO PIC-
tures as the next logical step. Certainly, if wires could carry pictures
so could radio waves. In this conviction, experimenters accepted the
challenge. They studied the older art of phototelegraphy, and called
the new art radio facsimile or radiophoto.

The telegraph and telephone had long ago inspired technicians to
send and receive pictures and printed matter over wires. In fact, such
experiments are almost as old as the telegraph; in 1842, Alexander
Bain, an English physicist, first produced a device to transmit pictures
over electric wires.

PICTURE SIGNALING

Arthur Korn,[1] a German pioneer in electrical transmission of pictures by wire, referred to a system he developed, as "seeing by wire."
He wrapped a photo-film around a revolving glass drum across which
a pencil-like beam of light traversed. The light ray regulated by the
lights and shadows of the picture was caught by a prism and focused
upon a light-sensitive selenium cell, connected with a battery. With

[1] In 1939, Professor Korn came to the United States and taught in the
electrical engineering department at Stevens Institute of Technology, Hoboken,
New Jersey.

this system, in 1904, Korn sent wire-photos over a telephone line from Munich to Nuremberg, 600 miles. His first wire-photos from the Continent to England in 1907 were heralded as a sensation.

Korn, fascinated by the possibilities of "picture signaling" by radio, turned his efforts from wires to wavelengths. In 1922, he demonstrated "phototelegraphy"; the apparatus comprised a wireless receiver hooked up to a typewriter or other mechanical printer so modified that it printed various sized dots instead of the letters of the alphabet.

Pictures radioed by this method were described as "half-tone groups of dots." To illustrate how it worked, Korn wired a picture from Centecello, near Rome, to Berlin, and from there it was radioed across the Atlantic to the U.S. Navy radio station at Otter Cliffs, Bar Harbor, Maine. It required about 40 minutes for the dots to "paint" the picture. A month later, on June 11, 1922, a transatlantic radio-photo of Pope Pius XI appeared in *The New York World* to win acclaim as a miracle of modern science.

RANGER'S SYSTEM

Captain Richard H. Ranger, a radio engineer returned from France after World War I, specialized in development of radiophoto and facsimile equipment, if for no other reason than the Army was vitally interested in being able to transmit and receive maps, etc., by radio. In 1920, Ranger joined the staff of RCA Communications to concentrate on "picturizing" radio.[2]

Within a few years he was ready to demonstrate the Ranger system featuring "photoradiograms." Front-page headlines on December 1, 1924 announced "Radio Flashes Pictures Over Sea from London to New York in 6 Minutes." Dozens of pictures were intercepted, including a photograph of President Coolidge, Secretary of State Charles Evans Hughes, Dowager Queen Alexandra, the Prince of Wales, and many others, as well as diagrams and, quite appropriate

[2] In 1968, RCA facsimile circuits to 63 countries carried thousands of radio-photos. Photolex introduced between New York and London in 1963 became available to numerous countries around the world via satellite and coaxial cable circuits providing customer-to-customer transmission of graphic material and documents.

to the occasion, in large letters the Chinese proverb, "One picture is worth 10,000 words."

With that as his inspiration for achievement, Ranger continued his experiments, and in May 1925, he radioed facsimile messages, maps, and pictures from New York to Honolulu; in April 1926, the picture-gram of a check was sent from London to New York where it was honored and cashed. Many pictures, advertisements, and fashion designs were flashed across the sea so successfully that commercial service was inaugurated on April 30, 1926. It was the lead story in *The New York Times,* which also printed a front-page radiophoto taken at the Pilgrims' Dinner in London.

THE PICTURE PROCESS

It is interesting to follow a picture from New York to London: The picture is wrapped around a horizontal cylinder, or drum, on a scanning machine. It revolves at a constant speed and in synchronism with a cylinder revolving at precisely the same speed on a recorder in London. A short audible dash, or synchronizing pulse, is transmitted when each revolution of the cylinder is completed. That signal in London automatically synchronizes the recorder and it is then ready to "copy" the picture components at the instant the New York transmitter sends them.

While the transmitting cylinder revolves, a spot of light moves slowly along its length until the entire picture has been spirally scanned, line by line. The light reflected from the picture during the transverse scanning is collected in a light-sensitive photocell. Whites, grays, and blacks reflect different amounts of light, proportional in intensity to the gradations of the picture. For example, whites reflect a maximum of light; and blacks a minimum. The photocell converts the light values into electrical currents of proportional strength, corresponding exactly with the whites, grays, and blacks of the picture. These electrical currents are amplified and made to actuate the transmitter beamed toward London.

There, the machine is identical with the one in New York except that in front of the cylinder is a photographic recorder instead of a scanner, and instead of a picture, a photographic film or sensitized paper is wrapped around the cylinder, moving in precise step with the one in New York.

The impulses received from New York are amplified and caused to control a fine beam, or "pencil" of light, thus exposing the film in London to the light that actuates the photocell in New York. In that way, the picture in New York is exposed to the film in London line by line, in direct relation to the lights and shadows of the picture being scanned. As soon as the picture transmission is complete—in about 10 minutes—the film is developed by standard photographic methods in a darkroom, and is ready for delivery by messenger.

THE GATEWAY TO TELEVISION

Radio facsimile was heralded as the gateway to television. If still pictures could be radioed across the sea, then the next logical step would be pictures in motion.

Facsimile, in 1934, was appraised as the first major advance in the art of telegraphy in 100 years, since Samuel F. B. Morse invented the telegraph. No longer was it necessary to break down the message or document into hundreds of dots and dashes; pictures or print could be reproduced by light "brushes" that "paint" them line by line.

Radiotelegraphy, to keep pace, was challenged to speed the flow of messages. The channel space that once carried one message at a time was made to handle twenty or more simultaneously. International radio became a teletype operation. Messages punched out on tapes automatically key the transmitter, while at the receiving terminal automatic teletype machines print the messages on tape at the rate of about 60 words a minute. As a result, on the overseas circuits, teletype has virtually replaced dots and dashes as well as the radio operator. For pictures, drawings and other illustrative material, radiophoto, of course, has the upper hand, but for messages and news, high-speed teletype radio is an expanding service.

Telex (teleprinter exchange service) since its inception in 1950 enjoyed rapid growth while computers extended and made more automatic the magic of its customer-to-customer services. Each Telex subscriber uses a typewriter-like machine with a dialing unit for direct connection with other subscribers throughout the United States, Canada, and Mexico. Or by dialing one of the international communication companies, a subscriber in this country can communicate directly with many thousands of Telex subscribers in more than 150 countries. Computer-to-computer facilities provide direct international transmission and reception of an avalanche of data processed accurately and instantly. A further service "Hot-line" provides exclusive voice and high-speed data communication between major cities.

CHALLENGING IDEAS

While the use of radiophoto increased for delivery of pictures and other items related to international news, foreign trade and business, some began to wonder if the applications of radio facsimile were being overlooked on the domestic front.

As visionaries studied the possibilities they suggested the day might come when a facsimile unit would be attached to standard broadcast receivers. Then, during the night after regular program hours, a newspaper in the home would be printed and available at the breakfast table. Others envisaged a printed program of the daily radio broadcasts, and, of course, both the radio newspaper and program would provide a new source of advertising revenue.

Owen D. Young, a founder and first Chairman of the Board of RCA, was among the first to pose the problem of using a radio machine to print a radio-facsimile newspaper in the home.

"I want a great camera arranged with a lens in London," he said, "and a plate in New York, so that when the London *Times* goes on the street five hours earlier than the morning papers in New York, the twenty sheets may be held up singly in front of the lens and, with a click for each sheet, be transmitted to New York, so that I may find a copy of the London *Times* awaiting with the New York papers on my breakfast table."

A prominent radio engineer, who heard Mr. Young fling this challenge remarked, "It's a fine thing to have an imagination wholly unrestrained by any knowledge of fundamental facts!"

Nevertheless Charles Young, trained in electrical engineering and radio at Harvard, took up his father's challenge. He shifted his activities from the field of broadcasting to radio facsimile and specialized in it at RCA Laboratories, always hoping to develop apparatus that would zip a newspaper across the Atlantic.

ULTRAFAX

To gain speed and increase radio's capabilities of handling great volume of printed matter, RCA in 1947 announced Ultrafax, which blended television techniques with radio facsimile, photography and other adjuncts of radio such as ultra-high frequency relays. This new high-speed photographic process was said to have a potential for handling a million words a minute and be capable of transmitting 50,000-word novels from New York to San Francisco in 60 seconds. The Library of Congress, Washington, D.C., provided an appropriate stage for the first public demonstration of Ultrafax as a new system of graphic communications, on October 21, 1948.

Ultrafax derived its point-to-point speed from its carrier radio—186,000 miles a second. Its unprecedented capacity in handling graphic communications was shown by the fact that it could service such assorted material as letters and telegrams, as well as signatures, drawings, maps, magazine illustrations, sections of newspapers, cartoons, advertising layouts, and all sorts of business and financial documents.

It could also handle musical scores, manuscripts, stock market tables, weather maps, and messages written in foreign languages whether in Chinese, Hebrew, or Russian. It was estimated that a complete Sunday newspaper including special supplements and comic sections, would require but a minute to be flashed from New York to San Francisco. To provide proof, an entire volume of the novel *Gone With the Wind*—more than one thousand pages—was transmitted across Washington over a radio relay circuit in less than a minute and a half.

"The remarkable speed," explained an engineer, "results from three key factors: First, the great velocity of radio. Second, the ability of television to transform pages of information for transmission as television pictures at the rate of 30 a second. Third, high-speed film processing, or 'hot photography,' which delivers a single frame of film ready for printing or projection in 45 seconds, as compared with 40 minutes by conventional process."

In this ingenious combination there seemed to be a promise that future generations might have a radio mail system that would deliver mail in bulk across the hemisphere in a matter of minutes. Donald S. Bond and his engineering associates, who developed Ultrafax, declared that even the airmail with messages carried by jet planes traveling above the speed of sound would not be able to compete with messages transmitted over radio relay systems operating at the speed of light; and the mail would be automatically reproduced as typed, or handwritten.

Radio, it was predicted, "will dip into the mail bag." If a letter is worth the time required for dictation, for a stenographer to type, for re-reading and signing by the sender, plus a stamp, and several days or weeks for arrival in Australia, why would it not be worth even a dollar to flash it to the Antipodes for swift delivery and a quick answer? Similarly, in the future, thousands of messages and letters might take wing from city to city, from country to country.

Indeed, the Ultrafax engineering staff saw great hope for its future. They confessed that they might be ahead of their time, but perhaps the day would come when satellite signaling, or some other development would summon Ultrafax to the forefront. They remembered how the reflected ultra-shortwaves that Hertz and Marconi demonstrated, finally, years later, were harnessed to bring radar, microwave communication and even Ultrafax, into being.

Often in the history of communications the necessity and urgency for a new service has brought an invention into commercial operation. In fact, it has been remarked that recognition of the need for an invention is half the invention. And, if the invention, or system, is blessed with mass appeal, its chance for service on a large scale is much better. True, radiophoto is a practical service for which there is demand, but it does not enjoy the mass appeal of radio broadcasting and television.

One might wonder whether such a conclusion induced Captain Ranger to forsake the radiophotos to specialize in pipeless organs and other electronic instruments designed to make music out of radio squeals, hums, and howls. In any event he thought it was "a new horizon of power and beauty."

Still another pioneer in radio facsimile, John V. L. Hogan, became a crusader for tonal quality and realism in broadcasting. To achieve the goal he was among the first to operate an FM station in New York; he built WQXR, New York's first high fidelity station, as he too sought new horizons of power and beauty, which he felt existed in radio but had to be cultivated.

The radiophoto art, however, continued to advance and encouraged scientists and engineers to take on the arduous task of developing ways and means to open the gateway to television. Based upon entirely different principles and techniques, television called for new apparatus—cameras, picture screens and many other devices. To transmit and receive still pictures was relatively simple compared with the broadcast of pictures in motion and in color.

IV   Television

I had heard of thee by the hearing of the
ear: but now mine eye seeth thee.
—*Job* XLII.V

TELEVISION, OFFSPRING OF RADIO, BEGAN TO OPEN ITS EYES IN THE
twenties. The radio "gold rush" was over and the economic depression that struck in 1929 caused broadcast station owners to stop, look, and listen. Many of them, however, had been making money hand-over-fist and were not over-anxious to look, especially at television that would call for greater investment.

Nevertheless, visual broadcasting was the next logical step after radio had learned to talk and to sing. Scientists and engineers again were pioneering on a new frontier—in a field of science called television.

"Tele" they took from Greek—"at a distance." And in Latin the verb "video" meant "I see." The Caesars might have called it "tele-video"; Americans preferred "television."

It called for new wizardry to combine the magic of radio, optics and everything new that electronics could offer. Radio broadcasting now looked simple compared with this new challenge to supplement the microphone with eyes. Engineers looked into history for clues on how to do the trick and to see if any devices were available as a start. Just as wireless had provided the foundation for radio broadcasting now radio broadcasting was the foundation upon which television could be built.

TV PIONEERS

A corps of scientists and experimenters set out to tackle the problem, among them: Ernst F. W. Alexanderson, Herbert E. Ives, Lee deForest, John Logie Baird, Charles Francis Jenkins, Vladimir K. Zworykin, Philo T. Farnsworth, Allen B. DuMont, Ulisses A. Sanabria, Alfred N. Goldsmith, R. D. Kell, and many others.

Hope for early success appeared to be in a mechanical scanning disk, invented by Paul Nipkow of Germany in 1884. The pioneers turned to his spirally perforated disk which could cut up a scene or image into tiny fragments for broadcasting. At the receiver the big riddle was how to re-assemble the pieces to duplicate the original scene. Each line that was scanned had to be "painted" electrically in exact sequence or the picture would be distorted and its identity lost. Thousands and thousands of light dots flitted across the screen, but the eye was not fast enough to see them all; it caught sight of only the complete picture.

Nipkow's Disk   Since there was no wireless in the eighties, Nipkow, in 1933, was asked what prompted him to develop the scanning disk. He replied:[1]

"One winter evening I was cheered by receiving from the Post Office the loan of a genuine Bell telephone for two hours in my room, which served as a living and sleeping chamber, as well as a laboratory and workshop. I was astounded by the remarkable simplicity of the telephone. I constructed a microphone out of nails and was successful in transmitting noises and words from one attic to another. This experience was what started me thinking about the problem of television. This problem stayed with me from then on, even during lectures of Helmholtz and Slaby in Berlin and Charlottenburg. Thus a sort of mental training along this line was developing in me, and, finally, on Christmas Eve, 1883, the solution came to me:

"The general idea of television. And the details including the perforated spiral distributing disk which has become known as the Nipkow disk. The mental experiment was a complete success. The ideas of the invention were automatically at hand—as are all everyday ideas.

[1] Letter to the author, February 27, 1933.

"Did I at that time think about the scope and the future of the electric telescope, alias television? Hardly; we must remember that then the use of the telephone was only in its first stages. Yet, the idea of television over telephone wires appeared before me. But then Heinrich Hertz had not yet taught; Marconi had not yet wirelessed. How then could such farflung ideas have come to a modest student of philosophy?

"How glad I was at the appearance of electron tubes and photocells which soon filled up all the gaps in my sketches and produced wonderful successes at the wireless expositions.

"In my opinion, the Braun (cathode-ray) tubes are likely to win more supporters, but without detracting from the value of the Nipkow disk. The simple, plain and solid style of the disk will certainly always have its admirers, particularly as the synchronization no longer offers any difficulties."

To some extent Nipkow was right; his disk whirled television into operation over wires and through the air.

Television on the Way   The first reference to the word "television" in the American press was found in the *Kansas City Star*—January 30, 1910—in a headline "Television on the Way," featuring a story from La Rochelle, France, telling how a young French scientist, M. Georges Rignoux (aided by M. Fournier), had demonstrated "some very interesting experiments in television." Various letters placed before "the transmitting telephoto" instantly appeared on a screen in a nearby room. Then images of a bottle and a lead pencil were instantly and accurately transmitted.

Apparently, these pioneers aimed at television over wires for no mention was made of radio. It was predicted however, that "the day is very near when one can sit comfortably in his own room and not only listen to the voice of a friend miles away, but see him as distinctly as though the friend were sitting in a chair beside him. And from his palace a monarch or president can inaugurate some public exposition thousands of miles distant being both seen and heard by the assembled people."

And in conclusion it was stated by M. Fournier, "We hope soon to transmit the colors as well!"

Steps Toward Success   By 1922, C. Francis Jenkins, an American pioneer in television, predicted "motion pictures by radio in the

home and an entire opera some day may be shown in the home
without the hindrance of muddy roads." In 1923, Jenkins placed a
portrait of President Harding in a camera-like affair at the Naval
Radio Station, Washington, D.C., and it was picked out of the air
130 miles away atop the Evening Bulletin Building in Philadelphia.

Events that followed, as experimenters revealed their progress,
indicated television was on the way all right, but there was no
promise of an instrument that would soon put it alongside radio in the
home as a household utility. Many a link was missing, but the
pioneers, encouraged by results and confident of the future, intensified their efforts.

Herbert E. Ives, electro-optical research director of the Bell Laboratories, was the first to show a "radio camera" that would televise
outdoor scenes without the glare of artificial lights; that was in 1928.
A year later, television in color was demonstrated at the Bell
Laboratories over wire from one room to another with the pictures
very clear but only the size of a postage stamp.

E. F. W. Alexanderson attracted front-page attention in 1930
when he showed television pictures on a six- by seven-foot theatre
screen in Schenectady, New York. An engineer stepped within range
of the camera, smoked a cigarette, and it was news that even the
wisps of smoke were seen on the screen.

At the same time John Logie Baird was winning fame in London.
His scanning disk whirled the picture of a pretty girl into space and it
was seen in the S. S. "Berengaria," 1,000 miles at sea; that was in
1928. And the face of Mrs. Mia Howe televised in London was
reported seen that year in Hartsdale, New York, as the first transatlantic television.

While the radio audience was content with Amos 'n' Andy, "soap
operas," and a wide assortment of shows, the New York air was
pulsing with pictures sent out in the efforts of RCA-NBC to contribute to the development of the new science and art. Just as the
phonograph turntable revolved and revolved for the radiophone
pioneers, now Felix the Cat sat upon the turntable in front of
television "eyes" that scanned him as the experimenters watched and
endeavored to improve the pictures.

Continually, new "firsts" in television were announced. More and
more TV got into the news—to the extent that *Variety,* the weekly
journal of show business, wondered in 1937 why *The New York*

Times devoted so much space to television, particularly since there was so little public interest and, said *Variety,* "Most dopesters see television as remote and uncertain when it does arrive."

This was about the same time that a survey was made on behalf of a major radio network to determine the prospects for TV broadcasting. When the report was submitted, it held little hope that television programming would be profitable and recommended that the network think twice before going into it. Before the foreboding was routed to top officials, it was handed back to the pollsters with the advice that they had better take another look. They did, revised the report, and went along with management in favor of telecasting. Anyone who had observed the success of radio broadcasting could not help but foresee a bright future for television. On both sides of the Atlantic 'round and 'round went the scanning disks, but television didn't seem to be getting very close to the American home. It needed some drastically new approach to get on the right track.

Television Becomes Electronic Perhaps the cathode-ray tube held the clue.

Sir William Crookes, British chemist and physicist, invented the cathode-ray tube, which up to 1895 was generally looked upon as a scientific plaything. But that year, Roentgen discovered that a Crookes tube enclosed in a darkened box emitted mysterious rays which made outside fluorescent materials luminous; he called them X-rays.

When Karl Ferdinand Braun picked up the Crookes tube for experimentation he found that it lacked only one element to enable it to reproduce luminous pictures in motion; the stream of cathode rays was uncontrolled. For television the electrons would have to be controlled; Braun had the electron gun, deflecting plates, and fluorescent screen, but the "gun control" necessary to control the electron beam to "paint" pictures was missing. That remained for another scientist to supply, but Braun pointed the way and shared the Nobel Prize for physics with Marconi in 1909 for his development of the cathode-ray tube.

Vladimir Kosma Zworykin was convinced that he had the correct solution—an electronic pick-up device, or "eye" with sensitivity necessary to make it practical for high-definition scanning. He replaced the mechanical scanning disk with an all-electronic system of television.

Zworykin came to America from Russia. He had completed an electrical engineering course at the Technological Institute in Leningrad. There he met Boris Rosing, professor of physics, and a pioneer in recognizing the possibilities of cathode rays as a means of reproducing electrical images by electromagnetic scanning. A. A. Campbell-Swinton in England had a similar idea and, in 1911, in a lecture before the Roentgen Society he proposed cathode-ray scanning. The idea, however, did not seem to lend itself to translation into practice.

Zworykin later studied under Paul Langeven at the College de France in Paris, where he learned more about X-rays, optics and photo-electric cells. He came to the United States in 1919, went to work in the Westinghouse laboratory at Pittsburgh and in 1930 joined the RCA staff.

By 1923, Zworykin had invented a cathode-ray tube as a television "eye" for the transmitter. He named it "Iconoscope," "eikon" in Greek meaning image, and "skopein" to watch. Then, for reception, he developed a funnel-shaped tube with a fluorescent screen across the flat end. He named that "Kinescope," "kinema" meaning movement in Greek. He first demonstrated the principle in 1923, and publicly in 1929 at a meeting of the Institute of Engineers in Rochester, New York. No longer did television need a motor, a scanning disk or any other moving parts. It was all-electronic.

Eventually, with further research and engineering, Zworykin was confident that practical, foolproof, simplified television for the home would be a reality. His next job, however, was out of the scientific realm; he had to convince industrialists that he really had something—the right approach to television. He estimated it would cost about $100,000 to develop the idea. And he proved to be as good a salesman for his idea as he was its scientific mentor, for RCA spent $50 million on research and engineering before it produced and sold the first commercial TV receiver.

TV FEATURED AT WORLD'S FAIR

Television was introduced by RCA as ready for public use at the New York World's Fair in 1939. As a curtain-raiser, the opening of the Fair was televised, and President Roosevelt, speaking from "The Court of Peace," became the first Chief Executive to be seen on television.

Events moved swiftly after that gala day to establish record after record for television. While New York's teeming millions gazed from every possible vantage point for a glimpse of King George and Queen Elizabeth in an awe-inspiring scene which radio announcers confessed "beggared description," televiewers sat comfortably at home watching on 200 to 500 TV sets that presented close and more intimate pictures than the crowd at the World's Fair could see. Throughout the Metropolitan area, Connecticut, New Jersey and upstate at Schenectady, 150 miles across the Catskills, the audience was amazed by the clarity and realism of television.

It was observed on that day in June "the home became a balustrade overlooking the rotunda of the world."

Television became commercial on July 1, 1941, when the NBC broadcast the first commercial TV program—the time, temperature and weather—sponsored by Bulova Watch Company. Next in line, for 15 minutes each, were Lowell Thomas with the news, sponsored by Sun Oil Company; "Truth or Consequences," by Proctor & Gamble; and "Uncle Jim's Question Bee," by Lever Brothers.

History repeated itself in that World War II stopped television for civilian or commercial use in much the same way that World War I had treated radio-telephony and the hopes for a "radio music box." But in the laboratories television research and development were speeded for the military. When the conflict ended, the radio-television industry was in a better position to produce a home product. TV was now on the launching pad of a new industry. And while it was confronted with obstacles and problems, it missed many of the pitfalls and road blocks that beset radio broadcasting. There was no question as to who would pay for television. All it needed was "circulation"— an audience large enough to make it worthwhile for advertisers to sponsor programs. Now they had a great new medium for visual advertising and appeal; they could demonstrate products and services in motion. A new automobile, electric razor, cereal, gown, or pill could be shown in homes across the nation in a single evening.

POST-WAR TELEVISION

Encouraged by the wartime advances, the Federal Communications Commission on November 21, 1945, announced new rules and engineering standards for immediate resumption of commercial television operation. Design and production of TV receivers got underway and the first to reach the market in September 1946, featured a 10-inch picture. As the public demand increased so did the screen size, from 10 to 12 inches, to 17 and 21, to 24 and 27, but 21 inches proved to be popular and became more or less standard. Before long seven TV stations were operating from an antenna mast atop the Empire State Building in New York.

President Truman's inauguration in January 1949 was the first event of its kind to be televised across 14 states by 34 stations in 16 cities from Boston to St. Louis. TV "firsts" multiplied in sports, news, national affairs and entertainment. Radio entertainers flocked to the cameras—some were telegenically successful; others were more successful unseen at the microphone. No doubt TV was capturing many from the radio audience and evidence began to appear that radio had seen its heyday, and that its hope for survival rested in performing services that TV couldn't—for example, in automobiles and through portable and pocketsize sets. As a medium for quick dissemination of news radio was still potent in reaching every nook and corner of the world.

For the Kennedy-Johnson inaugural in 1961 there were about 45 million TV sets and it was estimated that 85 million persons looked in on one of the most thorough and interesting jobs TV ever did. For those who had color TV sets, the inaugural parade was resplendent.

COLOR TV

Color had challenged the restless scientists ever since the war, but the TV station owners and manufacturers were none too keen about going into something new so quickly. Wasn't black and white good enough? Why rush? The "rush" got under way in the late forties and continued through the decade of the fifties, not only in the laboratories and in the air, but in the courts and newspapers. Based upon experience in black and white TV, the Radio Corporation of America, after exhaustive tests with the old mechanical scanning disk—again dusted off and brought forward to whirl color TV into service—cast it aside, determined to develop an all-electronic system more practical for the home. The Columbia Broadcasting System, however, continued to champion the scanning disk.

The battle was on. CBS attracted considerable attention in Washington and elsewhere with its colorful demonstrations, while the all-electronic system was not yet ready to match it. Scientist and engineers in RCA Laboratories worked day and night to find the missing links. Finally, Alfred N. Goldsmith, associated with RCA as a consultant, applied for a patent on a color television system featuring a color tube, which he invented. That was in August 1944, and in July 1947, he filed another patent application covering further improvements. The tube reproduced color pictures and black and white as well. It was called "compatible." That meant the home would not need a separate TV set for color and another for black and white.

Numerous other elements necessary for the complete color system were similarly evolved by other scientists and engineers in RCA Laboratories until finally, with practical development of the color tube, a complete and operative all-electronic system became available. RCA was ready to go. It informed the FCC in August 1949 that a high-definition all-electronic color television system had been developed, operating on a 6-megacycle channel and completely compatible with the existing black and white TV system. Field tests and demonstrations proved its worth and won acclaim as an outstanding achievement in the science of electronic communications. Goldsmith's color tube was heralded as one of the great accomplishments of the age. Those who had championed color produced by the antique scanning disk, discontinued crusading for the mechanical system.

The radio-television industry in general was reluctant and slow to take to color; excuses and arguments for delay were reminiscent of the reaction by radio broadcasters and some radio manufacturers, when television was bidding for entrance to the home in the thirties. Encouraged by the FCC's approval of standards for compatible color television on December 17, 1953, RCA pushed on alone and spent an estimated $130 million in research and development to back up its faith in the all-electronic color. Evidence of color's value has been seen in the World Series, in football games such as the Army-Navy classic, and in many entertainment programs, as well as advertising. Color gives life and realism; it enhances the interest of the eye.

The perfection and performance of color TV attests to the fact that the word "impossible" is not to be found in the vocabulary of scientists and engineers. And it would long be remembered that the counsel for CBS in presenting the color case before the U.S. Supreme Court declared in reference to the all-electronic color system, ". . . Why, the Commission [FCC] held it cannot possibly work."

Whereupon, Justice Felix Frankfurter said, ". . . Never is an awfully big word in this court. . . . I know enough of the history of science to know that science is the achievement of the impossible."

V Radar

Backward or forward, it's just as far;
Out or in, the way's as narrow.
—Ibsen, *Peer Gynt.*

RADAR, AS A NEW AND SPECTACULAR ADVANCE IN THE SCIENCE OF electronic communications, was born of necessity during World War II.

Heralded as "television's first cousin" and the most versatile weapon developed during the war, radar evolved in secrecy and under strict military security. It had the misfortune news-wise of being released at the time the atomic bomb dropped on Hiroshima, so that first official detailed disclosure of radar's dramatic story was dwarfed and muffled by the atomic blast.

It took some time for radar to catch up in the news, and as it did the public became more interested in its mystery. In theory they were told, "It is simple to the point of austerity, but its practice is one of the most complex arts of communication." And it was pointed out that its applications were only beginning. As the use of radio for nationwide commercial broadcasting was not seen until several years after military use of the radio-telephone in World War I, now radar stood on the threshold of industrial development.

RADAR'S BASIC FEATURES

It was explained as detecting and ranging by radio. *RA*-radio; *D*-detection or direction finding; *A*-and; *R*-ranging. That makes up the word "RADAR," the same forward and backward.

Simple to the point of austerity—yes; merely an echo. Radar is based upon an old principle, generally dormant until there came a day when it was vitally needed. A radio wave in striking an object, whether a mountain, ship, airplane, building, or even the moon and planets, bounces back, or echoes on the radio receiver.

Add a cathode-ray tube at the receiver and radar sketches a picture of what it sees on the radarscope, which is to radar what a picture-tube is to television. Ships, planes, clouds, coastlines, "maps" of cities, and even flocks of birds come into view. In fact, radar might well be described as an offshoot of the cathode-ray tube using ultra-shortwaves as boomeranging radio beams; it makes clay pigeons of planes, ships, rockets, and satellites within its range.

It was no secret to scientists that radio waves could be reflected. Hertz discovered that in his earliest experiments. And it was remembered that Marconi, in his tests at Salisbury Plain in 1896, demonstrated reflection of the waves from one parabolic mirror to another, a quarter of a mile apart.

"That apparatus," recalled Marconi, "had certain properties which the government was anxious to test. They were rather afraid of the waves spreading all around, and they were very keen to see something that would send the electric waves in one direction only. It was for that reason these experiments were carried out."[1]

Years later, in June 1922, he stepped out upon the stage in the auditorium of the Institute of Radio Engineers in New York to perform a fascinating demonstration that penetrated the future. He erected queer-looking skeletonized contraptions on both sides of the stage, and by means of a "baby wireless outfit" he showed how radio beams could be projected in a desired direction. A semi-circular, frame-like reflector at one end of the footlights projected the wave, while a horizontal rod on the opposite side of the stage caught the impulses, and instantly a clear-sounding note rang from the receiving instrument. Marconi paused in his lecture to emphasize a prediction that a revolution in radio was in the offing because microwaves, which obeyed the ordinary optical laws of reflection, were on the way.

---

[1] Testimony in injunction suit of the Marconi Wireless Telegraph Company of America against the National Electric Signaling Company in U.S. Circuit Court, Brooklyn, N.Y., October, 1913.

WAR RUSHED RADAR DEVELOPMENT

Rumors were whispered during World War II that a mysterious "ray" had been put into operation. Lord Beaverbrook, British Minister of State, in 1943, broadcast an urgent appeal for radio volunteers throughout the Empire and the United States. He alluded to "radio that destroys the enemy in the darkness and seeks him through the clouds," and added, "It is the radio that sends the avenging fighter to the place where he will meet the lurking enemy and bring him to destruction. . . ."

The British called it "radiolocation." Americans named it "radar." The Congressional Record referred to it as "the superweapon—the most revolutionary military device of this war. . . . It represents the greatest technical advance in warfare since the original evolution of the military uses of aircraft."

On both sides of the Atlantic new devices designed to make radar more effective were rushed to completion. The cavity magnetron oscillator, capable of generating microwaves far more powerful than any previously attained, was designed and developed at Birmingham University by John T. Randall, H. A. Boot, and F. G. Duke under the direction of Mark L. Oliphant. As an outstanding contribution of British science, the magnetron revolutionized radar giving it new power, range, and efficiency.

When the British sent a scientific mission headed by Sir Henry Tizard to America in the summer of 1940 they brought the magnetron as a new and powerful "heart" for radar. American scientists and engineers were quick to study it from all angles and, as a result, a series of vital developments were made at the Radiation Laboratory, Massachusetts Institute of Technology and at the Bell Telephone Laboratories.

In the United States, there was another important advance—the klystron, an ultra-high frequency microwave tube, developed by the Varian brothers, Russell and Sigurd. It projected a beam described as "straight as a sunbeam" for radar and aircraft guidance. All of these new developments as well as the cathode-ray tube and techniques of radio and television were mustered to make the miracle of radar a vital force in victory for the Allies.

Britain's Bastion   Early in 1936 the British Air Ministry had five radar stations on the east coast of England; in August 1937, 14 additional stations were authorized. And in 1938, when war was a probability, the radiolocation defense system was extended, so that at the time of Munich in September 1938, all available experimental equipment was rushed into operation. By March 1939, this defense system—Britain's "invisible bastion"—extended from Scotland to the Isle of Wight. More than $8 million had been spent on this work up to September 1938, and by March 1940, the total sum voted for expenditure was $36 million. Uninterrupted 24-hour radar watch was maintained on the North Sea approaches after Easter 1939 and when war broke out the radar sentinels were in continuous operation along the whole coastline of Great Britain. The battle of science was on.

When the Luftwaffe bombed England the German raiders encountered unexpected opposition. RAF fighter planes were out over the English Channel to give them battle and ward off the attacks. It was radar that was spotting the enemy and making it possible for the RAF to defend England; anti-aircraft guns as well as searchlights could be aimed and controlled by radar. Hundreds of German planes were destroyed in the aerial blitz.

"If radar had not prevented the enemy from getting over England by surprise," said Sir Stafford Cripps, Chairman of the British Radio Board, "I don't know where we would have been."

And the British, grateful for the role played by a Scotsman, Robert Alexander Watson-Watt in the development and application of radar, knighted him in 1942 as "a pioneer in radio location, who harnessed radar as a practical operational science."

Watson-Watt had been appointed superintendent of the Radio Department of Britain's National Physical Laboratory in 1933. Well aware of early observations of radio echoes he recalled how E. V. Appleton in 1924 measured the time taken by a radio signal to travel out and back from a reflecting atmospheric layer,[2] which he determined to be 60 miles up.

Early in 1935, Watson-Watt informed the Air Ministry that location of airplanes should be possible by radio. In May of that year, he and his staff set up a small laboratory at Orfordness on the eastern tip

---

[2] Kennelly-Heaviside layer.

of England, and it was there that tests were first conducted on what was to become known as radar. It was a great day in June 1935, when an aircraft 40 miles away was dimly tracked on the radarscope; and in March 1936, at a new experimental station at Bawdsey Manor, a high-flying Dutch airliner was detected 75 miles away.

Sir Robert later revealed a further wartime application of radar at sea: the decisive phase of the Battle of the Atlantic was won by fifty sets of radar equipment responsible for sinking more than eleven German submarines in the crucial period from March to June, 1943.

When the Eighth Air Force soared from its bases before daybreak on D Day—June 6, 1944—thick clouds concealed the English Channel. But the shore lines and the German gun positions were clear on the radarscopes, enabling the airmen to put down a 30-minute bomb barrage that momentarily paralyzed many of the Nazi defenses before the hour of the first infantry landing on the beach. Every bomb dropped was aimed by radar with effectiveness that made for a superb attack.

Historic Radar Advances   While the British on the frontline of attack were applying radar, major advances also were being made by scientists and engineers in the United States. The U.S. Army, Navy, and Air Force, as well as industrial and university laboratories were advancing radar as the most revolutionary military device of the war. British and United States scientists cooperated with great success in what constituted one of the most brilliant and important examples of Anglo-American collaboration. American ingenuity in electronic research and engineering coupled with unmatched skill in mass production hastened radar into a wartime service at a cost of many hundreds of millions of dollars.

The U.S. Navy traced its contribution to the early development of radar back to 1922, when A. Hoyt Taylor and Leo C. Young, at the Naval Research and Aeronautical Laboratory, recognized the principles of radar as they observed radio-echo signals from moving objects.

Basic research on radar techniques and apparatus was instituted by the Radio Corporation of America early in 1930. Two years later microwave equipment was used for a series of cooperative reflection tests with the U.S. Army Signal Corps near Sandy Hook on the New Jersey coast. These secretly conducted experiments proved that even with equipment in early stages of development, ships could be detected and located by means of microwaves.

Early in 1937, experimental radar apparatus was delivered to the Army Signal Corps for aircraft-location tests, and a year later equipment was installed on the U.S.S. "Texas" and the U.S.S. "New York."

The "secret" of radar first came into the news as far as its use by the U.S. Army and Navy was concerned, a few days after the Japanese attack on Pearl Harbor. It was announced that a Signal Corps sergeant had discovered what he thought was a large flight of planes East of North of Oahu at a distance of about 130 miles. A fleet of planes from the mainland of the United States was expected in that vicinity at the time, and the officer to whom the sergeant reported assumed that the planes were friendly and took no action. This incident taught a costly lesson: that radar was an instrument of science never to be ignored in modern warfare.

Radar stirred international scientific rivalry as it became increasingly evident that when the war ended, radar technology would be a new branch of communications and would perform new services on a world-wide as well as on a celestial scale. Its usefulness as a vital military weapon was universally recognized, and it was generally conceded that no war of the future could be won without radar.

After World War II, to protect North America from surprise attack across the polar region, the United States and Canada built a radar screen known as the DEW Line (Distant Early Warning) to detect and track approaching aircraft. A multi-billion dollar Ballistic Missile Early Warning System (BMEWS) also was erected across the Arctic to detect intercontinental missiles.

The DEW Line with 58 radar stations ashore and afloat stretched halfway around the world in 1963. By 1969, it had shrunk to 33 radars and was considered virtually obsolete, destined eventually to be replaced by an Airborne Warning and Control System (AWACS) featuring radar-equipped jets flying at high altitudes.

At Patrick Air Force Base, Cape Kennedy, Florida, and at other principal military ranges, radar traces the path of space-bound missiles and vehicles so that their performance can be evaluated instantly by digital computers operating to an accuracy of less than 2 inches error per mile. At the launching site, radar begins tracking the missile. And as it moves down range, for example, toward Ascension Island, 5,000 miles away, other radar stations on ships and islands follow the flight and flash a running story of performance.

Radar also is used to guide missiles. Radar on the ground "locks" on a target in the air and the missile travels to the target along the radar beam guided by its own electronic "brain"; this is called a beam-riding missile.

A compact lightweight electronic fire-control radar system has been developed for fast combat planes, such as the F-104 "Star-fighter" of the U.S. Air Force. Pilot reaction time in modern aerial combat could not suffice, and control of the aircraft must be supplemented by electronics. Radar-computer systems supply a continuous flow of information about target position in terms of range and rate of closing the gap between the plane and target. Fire control radar tells the pilot where to aim and when to fire, whatever the armament—guns, rockets, or missiles.

These same radar techniques are in service on warships with guns and rockets aimed and fired by radar to hit targets in the darkness at extreme ranges, representing the automatic power of push-button warfare. It was revealed during World War II that radar is endowed with unsurpassed accuracy in sharpshooting across miles of water in the night or through the thickest fog. For example, when H.M.S. "Duke of York" was seeking out the German battleship "Scharn-horst," it relied entirely on radar. The "Duke of York" first picked up the enemy at the extreme range of 45,500 yards (22.75 miles) and closed to 12,000 yards before firing, and the "Scharnhorst" was apparently unaware of the British battleship's presence until the first salvo was fired.

It is radar also that provides the main flow of information on which the missile control operators work—other sources being radio telemetry and optical instruments. Newly developed radars are so accurate that it is estimated that from a distance of 88 miles they could call "fair" or "foul" on a baseball hit at Yankee Stadium.

To accomplish all this, large and small rotating radar dish antennas scan the skies in the same way that the early radar antennas peered through the war-inflamed skies over the English Channel to spot hostile planes and buzz-bombs in flight toward the British Isles.

Wartime development of a radar Ground Control Approach system

(G.C.A.) for safe landing of aircraft under all weather and traffic conditions was heralded as "one of the greatest achievements in aviation in America," for there was dramatic evidence that during the war it had saved many lives as well as equipment valued at millions of dollars. Inherent advantages of military radar encouraged great efforts in development of precision radar, which has proceeded continuously since the end of World War II.

In the postwar period, the airlines looked to radar as a new instrument of safety and navigation. Soon airports were equipped for radar guidance in the thickest of weather, and its use in monitoring direction and altitude of landing planes became standard on instrument runways. As a double check, the pilot may avail himself of the Ground Control Approach system. The operator on the ground watches the plane on radar and informs the pilot by radio whether the plane is on course, high or low, right or left.

ALL-WEATHER EYES

Planes are equipped with all-weather radar, enabling the pilot to "see" and avoid storm formations up to 150 miles ahead. He plans his flight path accordingly, increasing passenger comfort and arrival-time dependability. Mounted in the nose of the plane is a radio transmitter-receiver device. Ultra-high frequency waves are projected in a narrow beam and when they strike clouds laden with precipitation they bounce back and are picked up by the receiver which visually indicates the distance, direction and extent of the storm.

Supplementing the plane's weather radar, there are ground weather radar stations in use by airlines and other members of the air transport industry. Radar clearly indicates the storm cores and areas of heaviest rainfall and can penetrate intervening rainfall and "see" storms that may exist beyond the rainfall area. Range, direction, relative intensity, and movement of the storms can be "seen" over an area of approximately 30,000 square miles. The ground radar apparatus and antenna are housed in a fiberglass beehive-like cover, or radome, which provides protection for the electronic elements and the antenna that sweeps the skies on weather surveillance.

The U.S. Weather Bureau atop Radio City, New York, is equipped with a "Stormfinder" radar designed to "see" weather 250 miles away; its big dish antenna scans a 200,000-square mile area to help the weather man forecast based on what he sees on the radarscope.

The U.S. Weather Bureau operates a network of 47 weather-surveillance radar stations along the Atlantic Coast and the Gulf of Mexico, in the Midwest tornado belt, and several on the Pacific Coast. Plans call for 56 such stations by 1972. All these modern weather "eyes" can penetrate massive cloud, rain, and snow centers, distinguish tornado funnels and detect hurricanes. The Bureau also has access to information from U.S. Navy and Air Force weather radars as well as F.A.A. air traffic radars in the Rocky Mountain area. Another application of radar to aviation is found in an altimeter designed to provide continuous and instantaneous reading of terrain clearance with a high degree of accuracy over land or sea. Radar signals projected toward the earth are reflected back to the aircraft, where an indicator converts the echoes so that the distance they travel can be read instantaneously in feet on a scale on the indicator face.

Experiments have been conducted in the application of color television techniques to radar in which the picture on the radarscope is converted into a TV picture in the control room. On a color television tube targets are portrayed in different colors; for example, planes under surveillance are one color and planes locked in as active targets are another.

SEMICONDUCTORS AID ADVANCE

Now transistors (semiconductors) with inherent long life, low power consumption and virtual freedom from effects of shock and vibration are finding a wide range of electronic applications in aviation-radar. They are leading to more miniaturized and lighter equipment. Miniaturization is heralded as more than a trend in current design—it is a practice built on the keystone of semiconductors (transistors and diodes). These devices virtually eliminate the heat problem in electronic instruments and decrease weight penalties imposed upon modern aircraft by expanding electronic requirements. It is estimated that semiconductors will populate at least 50 per cent of the sockets in airborne communication systems and that the trend will continue at an accelerated pace.

The need for many of the new developments traces back to the radio echo, harnessed through painstaking scientific research and engineering over a period of many years, by many men in many countries. Through their contributions to the advance of radar they achieved a new dimension in the science of electronic communications. And the applications of radar, empowered by the laser and other inventions, show every promise of extending its usefulness as man continues in the conquest of space.

WHO INVENTED RADAR?

Who invented radar? Who is its Marconi?

A. Hoyt Taylor, radar pioneer at the U.S. Naval Research Laboratory, pointed out that radar is not an invention. He explained that it is a development, just as radio broadcasting and television are developments in which many persons participated and contributed. And he added, "As one of my British friends remarked, there is plenty of credit in radar to go around."

Indeed, radar is a technological composite representing scientific teamwork at its best. Scientists, engineers and research men of various nations contributed to its evolution; it was a timely assembly of numerous achievements which all fit together to complete the jigsaw puzzle presented by radar at its inception when time and events summoned it out of the laboratory into practical service. It is a story of unceasing human endeavor that goes back to the epochal electromagnetic experiments of Hertz. It weaves through almost every ensuing development from Thomson's discovery of the electron to invention of the cathode-ray tube and a multiplicity of other electronic devices, as well as long years of study of electromagnetic waves, particularly their reflecting phenomena.

STRANGE ECHOES IN SPACE

It was in February 1928, that Jorgen Hals, a radio engineer at Oslo, reported to Carl Stormer, a Norwegian physicist, that while listening to shortwave broadcasts from Eindhoven, Holland, he heard two echoes of each signal. One, he figured, had encircled the earth, while the other of mysterious origin, was picked up about three seconds later. From where did the delayed echoes come? The nearest object was the moon. Radio waves could go there and reflect back in about 2.5 seconds; therefore, Stormer calculated, the long-distance echoes must have been reflected by an object beyond the orbit of the moon. In 1929, Hals heard the longest echo to date, 4 minutes 20 seconds, and estimated the double journey 20 million miles. His observations indicated that distances up to 48 million miles might be covered.

Carr V. Van Anda, managing editor of *The Times,* always on watch for new clues in science, read about the Stormer-Hals observations in *Nature,* and passed the item along to the radio editor: "These radio echoes," said Van Anda, "appear to be an important discovery and real news; it may lead to some new developments in communications. I suggest you look into it and follow it."

Van Anda was right. The echoes revealed that certain radio waves, especially those of very low frequency, penetrate the ionosphere, as well as the Heaviside layer, and travel into interstellar space. They signaled new possibilities in long-distance communication.

Anything like that fascinated Van Anda, for he had an overpowering interest in science and mathematics, especially in cosmogony and astronomy. The Einstein theory to him was news; and he explained it with such lucidity and thoroughness that even Einstein was astounded. There was nothing new in science so faint or feeble that it could escape Van Anda's "nose for news" and appraisal by his brilliant intellect—even a radio echo that whispered from space.

ELECTRONIC EXPLORATION OF SPACE?

Marconi, greatly impressed by the radio echoes, hinted at the possibility of their use in electric exploration of space. Addressing the Italian Society for the Advancement of Science on September 11, 1930, at Trento, Italy, he expressed belief that radio waves travel long distances, even millions of miles beyond the earth's atmospheric layer.

"Why not?" he asked, "since light and heat waves reach us from the sun, penetrating the atmospheric layer. Radio waves are reflected by bands of ions outside the magnetic field of the earth, sometimes at a distance from the earth of 25 million miles.

"Layers capable of reflecting electric waves exist at heights varying with the hour of the day and the season of the year. These layers also are influenced by the effects of light, by electric and magnetic activity of the sun, and by other causes yet unknown.

"Radio engineers should therefore keep in touch with the work of meteorologists and astronomers, but it is equally useful for the latter to keep in touch with the former, owing to the powerful means which modern developments have placed at the disposal of radio stations. Wireless echoes are among the most fascinating of phenomena, capable of disclosing the most useful facts."

FIRING THE IMAGINATION

With radio akin to light and traveling at the same velocity, why shouldn't it be able to reach out to the moon, sun and planets? Those who called such communication nonsense had cause to revise their thinking on January 10, 1946, when a radar signal beamed at the moon from Evans Signal Laboratory, Belmar, New Jersey, echoed back in 2.4 seconds; radar had traversed 478,000 miles in a round trip to the moon.

"The imagination is fired!" exclaimed Waldemar Kaempffert, science editor of *The New York Times*. "Space dwindles—even astronomical space. Man sends a radio feeler and actually touches the moon."

That lunar "peep" was a signal in the ears of Army radar men that the day would come when man would reach out and touch the planets; it stirred speculation on interplanetary communication and space travel. They foresaw the time when high-speed missiles would be guided by radar and electronics with such precision that the Arctic and Antarctic, the Atlantic and the Pacific, would be no more effective in preventing attack than was the Delaware River when Washington crossed it.

Since the moon first peeped, scientists have reported "hearing" meteors and echoes off the corona of the sun. They explain that when radar beams strike the pebbles of cosmic substances, they are reflected and cause brief whistles in earphones attached to sensitive receivers. This they declare to be sufficient testimony to substantiate prediction that the sky is the limit for radar. It would echo from Mars in about 6 minutes; from Venus in 4.5.

Thus, while science soars to new altitudes, the men of radio science plunge down the wavelength scale—or up the frequency scale—always approaching closer and closer to the infrared, ultraviolet ray, the X-ray and gamma rays of radium. That's light; and Nikola Tesla declared, "Light can be nothing more than a sound wave in space; and the shorter the waves the more penetrative they will be."[3]

Evidence that modern scientists recognize the significance of Tesla's statement is found in the development of "coherent light radar" in which an optical laser makes it possible to use narrow beams of light instead of microwaves. Engineers tell how this system enables them to project electromagnetic energy at a precise frequency 100,000 times higher than conventional radar frequencies. By this high concentration of energy in an incredibly thin beam, radar is given new and amazing powers of discrimination as well as extended range ideal for use in outer space.

Experimenters report that they have achieved results with a laser radar using a 4-inch dish antenna that would require a 6-foot dish for conventional microwave radar to equal. This system with its added features of compactness, light weight, and low power requirements has led to the prediction that a satellite so equipped could survey vast areas of space, tracking other satellites, space vehicles, and other distant objects. Therefore, while regular radar "sees" by microwaves, this new form of radar may be expected to probe space with high-intensity, sliver-like beams of light, described by Tesla as "nothing more than a sound wave in space."

[3] Interviewed by the author, April 1934.

VI   Space Age Communications

> I could be bounded in a nutshell and
> count myself a king of infinite space . . .
> —*Hamlet*

WITH THE VIVIDNESS OF A LIGHTNING FLASH, A NEW ERA IN ORBITAL
flight and space communications opened in 1961, when Major Yuri
A. Gagarin, of the Soviet Union, and Commander Alan B. Shepard,
Jr., of the United States, were rocketed into history as the first two
men in space.

Heralded as a cosmonaut, Major Gagarin, traveling more than
17,000 miles an hour, made a single spin around the globe on April
12, in a 108-minute flight at altitudes ranging from 109 to 188 miles.
The five-ton radio-controlled spaceship named Vostok (*East*), was
equipped with two-way radio transmitters operating on short and
ultra-shortwaves, enabling the twenty-seven-year-old spaceman to
send messages, while ground observers watched him in his cabin by
television.

In contrast to the secrecy that shrouded Gagarin's flight, many
millions eyewitnessed the drama on television and listened to the
radio, when Commander Shepard was rocketed across the fringes of
space on May 6. They saw the giant Redstone booster and heard its
thunder as it hurled the 2,300-pound Mercury capsule Freedom 7
into the sky from Cape Kennedy (then Canaveral) on the Florida
east coast at a speed of 4,500 miles an hour to an altitude of 115
miles and down a range of 302 miles in 15 minutes.[1]

[1] Cape Canaveral was renamed Cape Kennedy by President Lyndon B. Johnson, who on November 28, 1963, also announced that the United States space
facility at that location henceforth would be known as the John F. Kennedy

In the annals of broadcasting, probably no program ever stirred such excitement and drama, as an entire nation held its breath, watched, listened, and prayed. So vivid was the TV-radio portrayal that Americans everywhere in homes, schools, automobiles, offices, stores, factories, and streets were captured by a sense of personal participation.

For those who watched and listened, electronics provided an extraordinary insight into the Space Age. Science, organized with precision and perfection, revealed through television and radio the tremendous role of electronics in the conquest of space. What was the astronaut's pulse, respiration, temperature; what was the exact altitude of the capsule—speed, pitch, roll, and yaw? Electronic communications had the answers via radio, radar, and telemetry as doctors in the control blockhouse hovered over the instruments. Range officers kept their eyes fixed on an electronic status board on which lights traced the rocket's path and indicated the capsule's point of impact—"right on the button" as planned. It was perfect *astrogation*—navigation in space!

Throughout the flight, Commander Shepard radioed to the control center at the Cape, where another astronaut joined in two-way conversation as the spaceman calmly and methodically reported into a microphone inside his visor helmet. Script writers never wrote such an exciting drama, packed with tension and suspense, as that enacted by Commander Shepard in memorable words, "Everything A-OK," and "What a beautiful view!" as he scanned 800 miles through the periscope at the highest point of his flight. And finally his comment, "Boy, what a ride!" The commentator's dramatic account from the deck of the aircraft carrier "Lake Champlain" climaxed an unforgettable hour on radio and television. He glimpsed a mere speck in the sky, curving downward with its heat shield glowing red from friction with the earth's atmosphere; then he saw the big parachute open and float down to splash the historic capsule on the surface of the sea and the astronaut climb out to be picked up by a helicopter.

Radio-electronics had humanized, under the full glare of publicity,

---

Space Center. It is operated by the Department of Defense and the National Aeronautics and Space Administration (NASA) which was created by the National Aeronautics and Space Act that became law on July 29, 1958. Cape Kennedy is the site of the nation's No. 1 station of the Atlantic missile range, also the focal point of Project Apollo. The nation's orbital flights and satellites mentioned in this book were launched from the Cape unless otherwise specified.

a great scientific achievement and portrayed its magnitude as well as the genius and skills of scientists, engineers, and astro-electronic experts. Thousands of workers and thousands of companies contributed to the feat—a masterful combination and culmination of everything the science and technology of radio-electronics had to offer.

A RECORD-BREAKING NETWORK

The 18-station global tracking communications network of the Mercury project was described as "the largest peacetime construction job ever accomplished simultaneously around the world." Eleven of the installations were equipped with long-range precision radar; fourteen, with telemetry and voice communication facilities; and six had additional equipment to control the capsule and return it to earth if necessary. The tracking system won recognition as "the most advanced and powerful computer-communications system ever developed to compute and predict the flight of spacecraft from launch through orbit to impact." A ground communications system linked all stations in the network with the computing and control centers. Two ships, specially adapted, were stationed in the Atlantic between Bermuda and the Canary Islands, and in the Indian Ocean.

Every major development in electronic communications—past and present—seemed to be called upon to play a part. From the art of wireless came the channels of communication; from the radiophone and broadcasting came the microphone, etc., while radar and television provided watchful "eyes" so that the flying capsule was really never out of sight. Electronically, Shepard's colleagues were all around him. Seldom, if ever, had modern communications drawn so extensively from such a number of sources to perform its wonder.

When the commander got back to earth, radio and television were there, too, so that all the nation could see, applaud, and participate in the celebration. Millions saw him land by helicopter on the White House lawn to be honored by President Kennedy; they saw him greeted at the Capitol and joined in his hour-long press conference. They heard him describe the sensations of the flight and how it felt to be propelled in a capsule by a 78,000-pound thrust booster. As all America was introduced to six fellow astronauts on the platform with Commander Shepard, it was evident that new triumphs in space were in the offing.

LOOKING AHEAD AND UP

Inspired by what America had achieved in space and by its prospects as a spacefaring nation, President Kennedy, who watched television at the White House as Commander Shepard made history, declared, "We are going to make a substantially larger effort in space."

Within three weeks, the President went before Congress to deliver a special message recommending a sharp step-up in funds to meet national goals in space activity during 1961–1962, and an additional $7 billion to $9 billion in the next five years. In addition, he asked for $50 million to accelerate the use of satellites in world-wide communication and $75 million to aid development of a satellite system for global weather observation. All these and other major scientific projects were seen as vital for promoting the doctrine of freedom around the world and to strengthen "peace through space."

Said the President: ". . . Now is the time to take longer strides—time for a great new American enterprise—time for this nation to take a clearly leading role in space achievement which, in many ways, may hold the key to our future on earth. . . . Space is open to us now. And our eagerness to share its meaning is not governed by the efforts of others. We go into space because whatever mankind must undertake, free men must fully share. . . .

"First, I believe that this nation should commit itself to achieving the goal, before this decade is out, of landing a man on the moon and returning him safely to earth. No single space project in this period will be more impressive to mankind or more important for the long-range exploration of space. And none will be so difficult or expensive to accomplish. We propose to accelerate the development of appropriate lunar spacecraft. . . . No one can predict with certainty what the ultimate meaning will be of mastery of space. . . . I believe we should go to the moon."

GRISSOM IN LIBERTY BELL 7

On a significant flight, Captain Virgil I. Grissom, U.S. Air Force—America's second spaceman—was rocketed into the Florida sky on July 21, 1961, for a 16-minute, 302-mile suborbital path traveling

5,280 miles an hour and reaching an altitude of 118 miles above the Atlantic.

From the time the Mercury capsule Liberty Bell 7 rushed skyward until it hit the impact target area down the Atlantic range, a vast audience looked in and listened, relieved to hear the astronaut say he felt fine and all systems were operating successfully. Communications performed to perfection as radar and electronic computers watched and registered the capsule's trajectory. Telemetering was rated "excellent." Grissom was heard chatting by radio with Shepard, and even as the Liberty Bell re-entered the atmosphere radio faded only slightly and did not completely blackout. At 65,000 feet the voice from space was clear, and at 21,000 feet word flashed that the parachute had opened to check the capsule's speed of descent.

Radio from the aircraft carrier "Randolph" stationed in the target area supplied a running description of the scene as helicopters took off to pick up the airman and capsule from the sea. Suddenly communications vibrated with unexpected excitement when it was announced that after the splashdown the hatch had blown off and the capsule, shipping water, appeared unseaworthy. Seeing blue sky Grissom decided it was time to abandon ship! There were some hectic moments as radio told that the astronaut in his silvery space suit was swimming in the water; a helicopter lifted him up and landed him on the deck of the "Randolph." But the Liberty Bell, heavy with water, was too much for the helicopters to lift, and it sank in three miles of ocean. Grissom said later that he was sure he did not blow the hatch; electronic engineers thought that possibly a short circuit blew it prematurely. To Grissom on board the "Randolph" came a telephone call from President Kennedy, expressing congratulations on a flight which is logged in the annals of orbiting as "a creditable demonstration in space virtuosity."

Major Gherman S. Titov became the second Soviet cosmonaut on August 7, 1961, when he rode a five-ton spaceship, Vostok 2, seventeen times around the earth in 25 hours and 18 minutes, covering 435,000 miles as he traveled 18,000 miles an hour around the globe every 88 minutes. He broadcast by radiophone, and while over the Soviet Union his image relayed by television from the 160-mile high orbit provided the first direct look at a man in space.

The United States lofted a chimpanzee named Enos on a two-orbit mission around the earth on November 29, 1961. Encouraged by the performance and successful recovery of the chimpanzee and capsule in the Atlantic 250 miles south of Bermuda, plans were announced to put a man into orbit.

GLENN'S HISTORIC FLIGHT

U.S. Marine Lieutenant Colonel John H. Glenn, Jr., was selected to be the first astronaut to represent Uncle Sam on a flight-controlled orbital voyage. Rocketed into space on February 20, 1962, he made three 17,540-mile-an-hour orbits. All along the course Glenn maintained voice contact, reporting by radio while the most complex globe-encircling system of communications, electronics, telemetry, and radar ever assembled guided and tracked the Mercury capsule Friendship 7. It was recovered from the Atlantic 800 miles southeast of Florida, with the astronaut in excellent condition after four hours and fifty-six minutes in space.

The achievement was heralded as "a massive enterprise in American communications." Even Glenn's heartbeat, charted by radio-electronics, when amplified at tracking stations sounded like the rhythmic beat of a pile driver. James Reston of *The Times* called it "the greatest American ride since Paul Revere," and "news from Heaven all the way." Throughout the world a record-breaking audience on radio and TV shared every minute of the drama and tension.

"It's hard to beat a day in which you're permitted the luxury of seeing four sunsets," exclaimed Glenn.[2] "We're just probing the surface of the greatest advancement in man's knowledge of his surroundings that has ever been made, I feel. There are benefits to science across the board. Any major effort such as this results in research by so many different specialties that it's hard to even envision the benefits that will accrue in many fields.

"Knowledge begets knowledge. The more I see, the more impressed I am not with how much we know but with how tremendous the areas are that are as yet unexplored. Exploration, knowledge, and achievement are good only in so far as we apply them to our future actions. Progress never stops. We are now on the verge of a new era. . . . As our knowledge of this universe in which we live in-

---

[2] Addressing a joint meeting of Congress, February 26, 1962.

creases may God grant us the wisdom and guidance to use it wisely."

Scientists and medical experts foresaw the day when a computer-analyzer would act as an electronic "doctor" for astronauts in orbit. The instrument would be designed to monitor body functions and flash alarms if anything went wrong. For example, palpitation of the heart would switch on a system that would transmit the astronaut's electrocardiogram to earth. Body conditions of spacemen—temperature, respiration, electrical manifestations of the heart and brain—have been radioed to doctors on the earth; but when three or four astronauts travel aboard a spacecraft on journeys over a long period of time, surveillance by earth-stationed medical men would probably become impractical. Therefore, for prolonged space voyages each astronaut might have to be wired to his personal electronic "doctor" —a computer-analyzer that would monitor his body functions and sound an alarm if anything went wrong so that physicians at the control center could study the ailment and determine if a change in plans was necessary.

CARPENTER'S TRIP

Lieutenant Commander M. Scott Carpenter became the second American astronaut to orbit the earth on May 24, 1962, while a global chain of tracking stations monitored the flight and enabled a world-wide audience to follow the spacecraft Aurora 7. It was a day of excitement and suspense as Carpenter was heard conversing with the ground stations, but as the three-orbit trip ended the global audience suffered almost an hour of anxiety about his safe landing.

While coming down from orbit radio communication blacked out, as was expected, part way on the descent. But because the nose of the capsule was pointed too high at the time the retro-, or braking, rockets fired to retard speed and bring it out of orbit, the landing area was overshot by 250 miles. Ships and planes rushed to the scene in the Atlantic off Puerto Rico while radar scanned the sky to determine whether the capsule had survived the searing re-entry into the atmosphere. Almost an hour after Carpenter's voice was last heard a rescue plane reported sighting him bobbing in the ocean in a bright orange life raft. Three hours later he was picked up by a helicopter from the carrier "Intrepid," and later the destroyer "Pierce" retrieved the capsule. For the audience that looked and listened it had been a long, tense day, but with a happy ending.

RENDEZVOUS IN SPACE

A third Russian, Major Andrian G. Nikolayev was carried into orbit by Vostok 3 on August 11, 1962. By radio he communicated with ground stations, and he was seen on TV screens as the spacecraft made sixty-four orbits in 94 hours and 22 minutes. His image, picked up and recorded in Moscow, was relayed through Eurovision, the European television network—so that he was seen in Britain as well as other parts of Europe. The earth was encircled every 88.32 minutes at a speed of 18,000 miles an hour, 148 miles above the earth on a trip that covered 1,663,000 miles.

A day later (August 12), Lieutenant Colonel Pavel R. Popovich took off in Vostok 4, headed for a historic rendezvous in space with Vostok 3. From adjacent orbits, at times only four miles apart, the two cosmonauts talked with each other by radiophone and saw each other on television while millions of TV viewers watched them. Television cameras in the spaceships were beamed to Moscow, where the images were converted for telecast and fed to Eurovision; then on to London where the pictures were put on video tape or kinescope film and flown to New York.

Popovich made 48 orbits in 70 hours, 57 minutes, having covered 1,247,000 miles by the time the two spaceships landed, within six minutes of each other, in the Kazakhstan desert, one after four and the other after three days in the weightless realm of space. The cosmonauts, greeted as "the heavenly twins," had established new records in communications as well as in flight.

"We have a long way to go in this space race," said President Kennedy. "We started late. But this is the new ocean, and I believe the United States must sail on it and be in a position second to none."

SCHIRRA MAKES SIX ORBITS

Commander Walter M. Schirra, Jr., in Mercury capsule Sigma 7, was lofted into space from the Florida pad on October 3, 1962. He was bound on six 88.5-minute trips around the world as millions of Americans watched the launching on TV and listened to the drama of it all on radio. And within forty minutes the Telstar communications satellite was in position over the Atlantic to relay taped pictures of the blast-off to Europe.

Throughout the day a world-wide audience followed him as he traveled 160,000 miles in nine hours and fourteen minutes. He made a pinpoint landing in the Pacific about 330 miles northeast of Midway Island, where the U.S.S. "Kearsarge" was stationed. There was no overshoot this time; the capsule, lowered by parachute, splashed on the water just four miles from the "Kearsarge"; and within minutes the astronaut in the bellshaped capsule, which he called "a sweet little bird," was lifted to the deck of the Navy ship.

Space officials described the flight as "almost flawless," and that applied to the global communications system's performance in every aspect. Here was a man traveling 17,560 miles an hour on an orbit ranging from 100 to 176 miles above the earth; yet medical experts at the control center in Florida were aware at all times of his exact heart beat, respiration, blood pressure, body temperature—all communicated by radio.

Schirra's voice, strong and clear, was heard in communication with the global tracking stations. And as he dashed past Hawaii he called out "Aloha." When informed that it was "Go" for the sixth orbit he exclaimed, "Hallelujah!"

Prior to the descent, the radio-TV audience was warned not to be alarmed, but to anticipate about four minutes of silence at the time of re-entry. It was explained that an ionization blackout of radio from the capsule would be caused by an ionized sheath of plasma (hot ionized gases) that clings to a spacecraft and repels radio waves as it passes through the ionosphere. But the sheath would be burned off when the craft entered the earth's atmosphere. Then the blackout would end—in much the same way that an automobile radio blacks out as it passes under a steel bridge, then quickly comes back into action as soon as the car gets into the open.

Schirra had named his spacecraft Sigma 7, the Greek letter sigma being a mathematical term signifying summation. But his orbital triumphs were only a prelude to voyages far out on the cosmic sea.

COOPER'S "GO" FOR 22 ORBITS

Speeding around the world 17,157 miles an hour in an orbit that ranged from 100.2 to 165.8 miles high, Major L. Gordon Cooper, Jr., of the U.S. Air Force, in a magnificently executed 22-orbit flight, showed that Americans were on their way to the moon and beyond. Blasted into space on May 15, 1963, he was off on a 34-hour, 589,050-mile journey that dramatically demonstrated how "aviation" had progressed since 1927 when Charles A. Lindbergh flew non-stop 3,900 miles from New York to Paris in 33.5 hours. Not a word was heard from Lindbergh's radioless plane "Spirit of St. Louis" after he disappeared over the sea until he was sighted over England.

No such silence shrouded Cooper's Faith 7. Radio, radar, telemetry, and television never lost "sight" of him. A constant vigil was maintained as the Mercury capsule encircled the earth every 93 minutes high above more than 100 countries; two-way communication with the tracking stations was continual. Ideal weather enabled the astronaut to conduct scientific experiments and to take pictures. On the first orbit he turned on the TV transmitter for reception on the Canary Islands. Later he checked to determine whether telemetric information could be sent over the television transmitter, and it proved to be a good TM as well as TV transmitter.

Even when the major drew curtains over the window of the capsule to sleep for five to six hours, electronic eyes and ears watched over him. In fact, when instruments indicated that his pulse rate shot up suddenly and briefly from 60 to 100, doctors attributed it to a lively dream. Throughout the slumber, ground stations received automatic telemetered reports on his heart action, breathing, pulse, and temperature. And the major reported that during one nap he was awakened by "some very nice music" followed by a language foreign to him, but he thought it might have been a news broadcast. Proof, indeed, that the cosmic sea is not "an ocean of dreams without a sound."

Many millions of TV spectators and millions of listeners abetted by portable transistor radios all around the world followed the flight from the dramatic vault into space to the splashdown on the Pacific off the island of Midway.

Apprehension and the dramatic element of suspense came into the TV and radio when, nearing the final orbit, Major Cooper reported a green light flashing on the instrument panel long before it should to indicate that the capsule had begun to re-enter the atmosphere. The astronaut and communications men on earth had some busy moments to determine how serious the premature warning was. The automatic re-entry system had developed trouble; so Cooper proceeded on manual control, while astronaut John H. Glenn, Jr., on board the tracking ship "Coastal Sentry," 275 miles southwest of Japan, communicated with Cooper and counted him down on the retrofire. The major pressed the button to fire the first of his retrorockets, and the descent was underway.

Communication from the downcoming Faith 7 was momentarily lost during the blackout period when the craft encountered the tremendous heat of re-entry. Soon, however, Cooper's voice was heard again; the capsule's big orange-and-white parachute had opened, and radio from the U.S.S. "Kearsarge" near Midway Island described the splashdown on the sea about 7,000 yards away. Helicopters circled, and frogmen attached flotation gear to the capsule. Shortly, Cooper was on deck. By unerring guidance he had made a bull's-eye landing, for he had fired the retrorockets "right on the old bazoo." Via radio-telephone President Kennedy extended congratulations on "a great flight." Hawaii's triumphant welcome, the arrival and press conference back at the Florida base, the Washington celebration, visit at the White House, and address before a joint session of Congress plus New York's tumultuous ticker-tape parade, all added to the TV spectacular, parts of which were seen throughout Europe via the communication satellites Telstar and Relay.

WOMAN IN SPACE

A month later, the Soviet Union orbited two more cosmonauts. Lieutenant Colonel Valery F. Bykovsky took to the skies in Vostok 5 on June 15, 1963; and two days later 26-year-old Junior Lieutenant Valentina V. Tereshkova, the first woman in space, was whirling around the globe in Vostok 6.

Colonel Bykovsky made 81 orbits, covering 2,046,000 miles in five days less 54 minutes in orbits ranging from 99 to 140 miles high.

Lieutenant Tereshkova orbited 48 times, traveling 1,240,000 miles in 70 hours, 50 minutes at altitudes ranging from 108 to 143 miles.

Identified on the radio as "Hawk" and "Seagull," both were expert communicators as they talked by radiophone and were seen on television. So clear was the TV that newspapers published pictures of them as they encircled the earth in nearby orbits, at one point about 3 miles apart. Extensive television coverage throughout the world, including use of Telstar 2 and Relay satellites, enabled a vast audience to see them as the Vostoks orbited every 88 minutes, traveling 18,000 miles an hour. The 28-year-old colonel performed by unbelting the harness of the pilot's couch so that he might be seen floating weightlessly upside down in the cabin. He was also seen eating; he held up his log book and again demonstrated weightlessness by letting a pencil and other objects float as "trifles light as air." His radio transmitted on 20.006 and 143.625 megacycles, while a signal transmitter operated on 19.948 megacycles. Telemetry reported the pulse beats, respiration, etc., of both cosmonauts to ground stations.

IMPOSING JOURNEYS

"I am sure that over coming years we can find a few hundred, or possibly a few thousand astronauts who would be glad to make exploratory journeys to the moon, to Venus and Mars," said Lee A. DuBridge, president of the California Institute of Technology.[3] ". . . Since the moon has no atmosphere at all and the atmospheres of Venus and Mars appear to contain no oxygen, the problems, even of survival, on such bases clearly present horrifying difficulties. To think of millions of people living under such circumstances clearly is getting close to the borders of insanity. . . .

"Within a relatively short time we shall be able to send spacecraft to the vicinity of the two nearest planets in our solar system—Venus

[3] Speaking at the Commonwealth Club, San Francisco, California, February 24, 1961. DuBridge was appointed science advisor to President Nixon in December, 1968.

and Mars. Only three or four months of travel, at presently attainable speeds, will bring us to the vicinity of these, our two nearest neighbors. In another ten years or so, our space capsules should be able to reach the vicinity of Jupiter. And it is certainly not out of the question of getting to the vicinity of Pluto, the most distant planet, although such a journey will take us more than two billion miles away from the earth and will certainly require many months, or even years, of travel time.

"From the astronomical point of view," continued DuBridge, "our human space travels are rather puny efforts; from the earthly point of view these travels are imposing journeys indeed. The entire solar system, a ball of space several billion miles in diameter surrounding our sun, populated with nine chunks of matter called planets and uncounted smaller pieces called asteroids filled with extremely sparse but measurable clouds of gas and dust, traversed by intense streams of all kinds of radiation coming from the sun and from the unknown reaches of outer space—all of this is an area of exploration presenting one of the greatest challenges in human history. . . . Our own solar system is plenty large enough to keep us engaged in space exploration and research for hundreds of years to come."

RUSSIAN SUCCESSES

The coming years to which DuBridge referred, even in their infancy, proved the accuracy of his vision as new names were added to the astronautical roster and millions of miles were charted on exploratory journeys far out into space.

The first three-man spaceship Voskhod (Sunrise) was orbited by the Soviet Union on October 12, 1964, at a perigee of 111 miles, apogee 254 miles. The crew, Colonel Vladimir M. Komarov, commander, Konstantin P. Feoktistov, a scientist, and Lieutenant Boris B. Yegorov, physician, were seen by television viewers in the Soviet Union as they conversed with ground observers by radio. After 16 trips around the globe the craft landed, having traveled 437,000 miles in 24 hours and 17 minutes.

Staging a "space spectacular" on March 18, 1965, the Soviets launched a 12,529-pound Voskhod 2, carrying Colonel Pavel I.

Balyayev, pilot, and Lieutenant Colonel Aleksei A. Leonov. Attached to a 5-yard rope lifeline Leonov became the first man to step out of a spacecraft for a ten-minute "walk" in space. After covering 447,000 miles in orbits ranging from 108 to 309 miles high the Voskhod landed, having made 17 trips around the earth in 26 hours. Transmissions on 17.365, 18.035, and 143.625 megacycles enabled viewers in Russia and Europe to watch Leonov somersault in space, while live telecasts also were made from the spacecraft's cabin.

"The target now before us is the moon," exclaimed a Soviet space official, "and we hope to reach it in the not distant future."

The space race was accelerating. So was the marathon to the moon.

Man's destiny in the Space Age prophesied by Konstantin E. Tsiolkovsky (1857–1935), known as the "father of Russian rocketry," is recorded on an obelisk over his grave at Kaluga: "Man will not stay on earth forever, but in the pursuit of light and space will first emerge timidly from the bounds of the atmosphere and then advance until he has conquered the whole circumsolar space."

A PIONEER MANEUVER

The Voskhod had no sooner landed when Major Virgil I. Grissom and Lieutenant Commander John W. Young took off on March 25, 1965, in Gemini 3 in which they accomplished the first maneuverable manned flight.[4] After three orbits in 4 hours and 54 minutes they splashed down in the Atlantic east of Bermuda, missing a planned landing near the aircraft carrier "Intrepid" by 50 miles. While the ship sped in their direction a Navy helicopter lifted the astronauts from the capsule and set them down on the deck of the "Intrepid," which an hour later retrieved the spacecraft.

Adding to the suspense at re-entry, an extra-long 12 minutes of radio-blackout silence ensued from the time the parachute opened until a search plane spotted the spaceship on the sea. During the three-orbit flight a vast audience watched on TV and listened to radio as the

---

[4] Gemini 1, an unmanned 7,000-pound capsule, was orbited by a Titan 2 rocket on April 8, 1964; apogee 204 miles, perigee 99.6 miles. Gemini 2, unmanned, made a 2,000-mile flight down the Caribbean on January 19, 1965.

Gemini went round-and-round, flying with its nose forward and backward, as well as upside down, and altering the flight path in various ways.

"The only thing wrong with it," said Commander Young, "was it didn't last long enough."

A COLORFUL BLAST-OFF

Gemini 4, a 7,800-pound spacecraft carrying Major James A. McDivitt and Major Edward H. White 2nd, was launched on June 3, 1965, bound on a 62-orbit, four-day flight ranging from 103 to 182 miles above the earth. Radio and television covered the blast-off, and for the first time many saw such an event in color. Europe looked in via the Early Bird satellite.

During the third orbit White went out the exit hatch on a 25-foot tether and floated in space for 20 minutes. With the spacecraft's radio circuit open, listeners could eavesdrop on the conversation between the two astronauts as well as instructions radioed to them from Ground Control. McDivitt's voice traveled by wire over the tether to White's helmet, and also to earth by radio. Sensors attached to the astronauts reported their pulse, temperature, and other body data.

After a flight of 97 hours, 58 minutes, Gemini splashed down in the Atlantic some 900 miles east of Florida where a helicopter from the carrier "Wasp" picked up the spacemen as well as the capsule. McDivitt and White proved themselves topnotch communicators; tracking was excellent throughout the voyage.

A 3.3 MILLION MILE JOURNEY

Piloted by Lieutenant Colonel Gordon Cooper, Jr., and Lieutenant Commander Charles Conrad, Jr., Gemini 5, a 7,000-pound spacecraft, was launched on August 21, 1965, bound on a 3.3-million-mile journey. After having made 120 circuits of the globe in 190 hours and 56 minutes, the Gemini splashed in the Atlantic 760 miles east of Florida. A helicopter hoisted the astronauts from the sea and landed them on the deck of the aircraft carrier "Champlain." Excellent communication was maintained throughout the flight, including frequent measurements of the spacemen's heartbeats, temperatures, and blood pressures. Radio and television enabled a countless audience to follow the flight from blast-off to recovery.

An innovation featured fast transmission of still pictures micro-waved from the "Champlain" to the NASA Manned Spacecraft Center at Houston, Texas, providing almost instantaneous visual confirmation of the astronauts' safety. Soon after re-entry, radar spotted the capsule and televiews of the radarscope were seen on TV screens throughout the country.

THE GEMINI SPIRIT OF 7–6

The flight of 7,000-pound Gemini 6 piloted by Captain Walter M. Schirra, Jr., and Major Thomas P. Stafford was canceled on October 25, 1965, about an hour before they were to be rocketed to chase an Agena rocket and undertake a rendezvous with it. The Agena mal-functioned and was lost out over the Atlantic six minutes after launch. Therefore, it was decided to hold Gemini 6 until Gemini 7 got into orbit in December when the two spacecraft could attempt a rendezvous.

The 8,069-pound Gemini 7 soared into orbit on December 4 be-ginning a 14-day journey destined to be record-breaking. Aboard were Air Force Lieutenant Colonel Frank Borman and Navy Com-mander James Lovell, Jr.[5] The lift-off was perfect. The Mission Con-trol Center radioed, "You're right down the slot!" Borman-Lovell in a matter of minutes were well on the way toward Africa with 206 circuits of the earth ahead of them. Scientific tests and observations as well as photography were planned to keep them busy aside from their navigational and communication tasks.

One experiment was to feature a 6-pound, camera-size, infrared laser transmitter designed to enable them to talk to earth over a coherent light beam. On the 105th revolution of the flight the laser was aimed at the tracking station on the island of Kauai, Hawaii. The signals appeared as faint blips on a tape-recording device, but were not strong enough to carry the voice. The tests, therefore, 17 in all,

[5] The Government adopted a policy in August 1965 to raise a military astronaut one rank on completion of a successful flight, but not beyond colonel in the Air Force, Army and Marines or captain in the Navy.

were reported "not a complete success"; unfavorable weather, rain, and ground equipment problems were blamed.

HISTORIC RENDEZVOUS

Eleven days after Gemini 7 was launched, the delayed Gemini 6 rocketed skyward for a rendezvous. Considering the vastness of space it seemed like a needle-in-a-haystack mission. Nevertheless, in about five hours, Gemini 6 searched out the sister ship by radar. Then a radar lock-on achievement brought them into rendezvous 185 miles above the earth, while a computer automatically helped them to maneuver and navigate with unprecedented precision. They sailed along nose-to-nose within 6 to 10 feet of each other while traveling about 5 miles a second. They photographed each other, and radio transmitters (3 watts) linked them in two-way conversation.

Radar was quick to track down No. 7 and put No. 6 on the right trajectory to catch up with it. When Gemini 6 was about three hours and ten minutes in flight, the two spacecraft were 275 miles apart. That was estimated to be the maximum range of the radar installations. By the time Gemini 6 was on its third orbit the radar lock-on was accomplished to be maintained throughout the rendezvous. The radar transmitter on board Gemini 6 locked on to the radar transponder of Gemini 7, which returned signals that were translated into position data by a computer aboard No. 6, then more than 200 miles behind. Borman radioed, "We are reading you loud and clear."

"See you soon," replied Schirra. "We will be up there shortly."

Mission accomplished, Gemini 6 peeled off from formation, and, having made 16 orbits, splashed in the Atlantic 700 miles southwest of Bermuda on the morning of December 16. No. 7 continued in orbit to complete its scheduled two weeks in flight. Schirra and Stafford landed Gemini 6 just 14.6 miles from the aircraft carrier "Wasp," about 2 miles off their theoretical target. For the first time the recovery of a spacecraft was televised live. A 10-kilowatt microwave TV transmitter beamed a fascinating panorama from the deck of the "Wasp" to the Early Bird satellite hovering some 22,300 miles above the South Atlantic, as it had since April 1963.

The pictures generally were of studio-brand clarity despite the fact that they traveled 46,000 miles from the "Wasp" to the Bird and down to the ground station at Andover, Maine, where the micro-waved images were fed to the regular television networks. It was a magnificent portrayal of television in its greatest natural role, *im-mediacy* sparkling with realism and excitement.

The long-peering TV lenses provided astonishing views of the floating spacecraft. Smoke from the marker bomb left no doubt as to the capsule's location on the waves. Navy frogmen were seen lashing the flotation collar in place as the helicopters hovered overhead. Another innovation, introduced at the Gemini 6 blast-off, featured an electron telescope developed at Boston University. It gave the TV audience spectacular long-range glimpses of the spacecraft when it was 400 miles from the launching pad and 98 miles up. Color televi-sion also came into play enhancing the studio-animated simulations and maps as well as prefilmed pictures of the capsules and the carrier "Wasp." Color injected thrilling realism into the blast-off scene and into the rocket flight in the clear blue sky.

RE-ENTRY OF GEMINI 7

Gemini 7 enacted another thrilling act in the tense drama of space on December 18 when it began the searing re-entry from 14 days of weightless environment. Borman and Lovell radioed everything in A-1 shape. The retrograde rockets fired automatically by computers were "right on nominal." Radar and the TV cameras on the "Wasp" began to scan the heavens and the sea. Helicopters were in the sky. Gemini 7 was on the way down, ending the longest journey in history, having traveled 5,716,000 million miles on 206 circuits of the earth in 330 hours and 35 minutes.

Shortly after the usual period of radio blackout, suspense lifted with word from the descending capsule that the main parachute had opened. Soon lookouts on the "Wasp," 12.6 miles away, reported that they could see the spacecraft; that the frogmen had dropped from the helicopter and attached the flotation collar, just 4 minutes after splashdown. It was only 7.6 miles from the predicted impact spot. The big carrier was closing in only about 4 miles away, 700 miles southwest of Bermuda in about the same spot where Gemini 6 had been hauled out of the sea.

Again Early Bird relayed extraordinary pictures. The TV audience was witnessing a spacecraft recovery by helicopter, which the astronauts elected. Millions of North Americans and Europeans saw the pilots hoisted up from the rubber raft onto which they had climbed. Focused off the bow the electronic eyes followed the helicopter approaching for touchdown on the carrier's runway just 31 minutes after splashdown. The ship's band struck up "Anchors Aweigh." The big rotary blades of the helicopter came to a stop. The hatch opened and out stepped Borman and Lovell with smiles as happy as any electronic camera had ever scanned. The ship's bell rang welcome aboard as the spacemen were saluted and applauded by the crew as they walked down the red carpet. Elated and alert they appeared to be in topnotch condition that matched their spirit—"the spirit of Gemini 7–6."

As the exultant spacemen entered the sickbay for medical checkup, the TV cameras switched back on deck to watch the crane lift Gemini 7 from the water. And to cap off the program, unsurpassed color pictures showed Gemini 7 traveling through space at 17,500 miles an hour as photographed from Gemini 6 while the two spacecraft were flying in rendezvous high above the Pacific.

Again television was at its best—a historic event, dramatic scenery, star actors, plenty of excitement, suspense, and action—all live as it happened as news and history were made! The TV audience was in the front row; there are no back row, off-side, or gallery seats in the television amphitheatre. No one on the "Wasp" had a better view of the show. Every TV viewer must have sensed that he too was on the flight deck to welcome the fliers, who, under the eyes of television, had literally landed on an international stage erected by modern communications.

Schirra and Stafford, Borman and Lovell, of the Gemini twins, not only made historic advances in man's journey into the solar system but established fantastic records in communications. Their superb skills as communicators matched their astronautic abilities. Precision and efficiency achieved through years of planning and training, supplemented by the remarkable performance of electronic instruments and all phases of communications, made these flights "look easy," as a space official remarked in summing up the triumph as the biggest milestone in space up to the end of 1965.

Extending congratulations to the spacemen and all who had anything to do with the feat, President Johnson said, "You have all moved us one step higher on the stairway to the moon."

GEMINI 8 SPACE DRAMA

Piloted by Major David R. Scott, U.S. Air Force, and Neil Armstrong, Gemini 8 was lofted on March 16, 1966. Its first assignment was to dock with a 27-foot Agena rocket launched ahead of it. Six hours later after a 105,000-mile chase, the astronauts guided by radar rendezvoused with the target and achieved a docking while both vehicles were traveling 17,295 miles an hour 185 miles above the earth.

Shortly thereafter, as startling as an SOS, came Armstrong's voice, "We consider this problem serious. We are toppling end over end." A smooth and successful space mission up to that point turned into a drama of life and death. Relaxation and jubilation at the Manned Spacecraft Center in Houston, Texas, that resulted from the docking accomplishment, turned into high tension and feverish activity.

While the Gemini was tumbling and spinning like a top over Southeast Asia, John Hodge, flight director at the Control Center, quickly decided to bring the astronauts down immediately—as soon as they could reach a landing site; he ordered an emergency splashdown. Some time later, an encouraging message, "We are regaining control of the spaceship slowly," lifted hope for a successful landing. The destroyer "Leonard F. Mason," about 60 miles away from the spot where the Gemini was expected to come down, was dispatched full-steam.

With extraordinary pilot skill, the splashdown was made in the Pacific 500 miles southeast of Okinawa. Meanwhile, an Air Force C-54 transport plane flew out, spotted the bobbing capsule, dropped an emergency raft, and parachuted frogmen who attached a flotation collar. Radio from the plane flashed the word that Scott and Armstrong looked good. Three hours later the destroyer "Mason" picked them up and retrieved the capsule.

Millions of Americans in tune with the space drama had followed the flight from the hour of blast-off, were thrilled by the news of the successful docking, and were standing by for Scott's scheduled two-hour walk in space. Throughout the day they had heard reports of a perfect mission, serene and successful. Then came the ominous announcement that Gemini 8 was in danger, tumbling out of control. Communications activity was greatly intensified over one of the most extensive networks ever put into action—from plane to plane, to ship, to shore, and to broadcasting stations everywhere. Paul Haney's voice from the Mission Control Center in Texas, announcing that the astronauts had been rescued, ended hours of suspense throughout the anxious world.

What had happened? After seven orbits and "a magnificent flight" for seven hours, as Armstrong described it, maneuverability of the Gemini was lost. A short circuit in the electrical system that controlled a thruster—a maneuvering unit—was blamed for the near disaster.

President Johnson, who watched the launch on TV, affirmed that the United States still intended "to land the first man on the surface of the moon in the decade of the sixties."

GEMINI 9'S "ALLIGATOR" TARGET

As a further significant advance toward that lunar goal, Gemini 9 piloted by Lieutenant Colonel Thomas P. Stafford, of the U.S. Air Force, and Lieutenant Commander Eugene A. Cernan, U.S. Navy, rocketed upward on June 3, 1966, bound on a three-day flight.

One of their primary assignments was to dock with an 11-foot target vehicle launched several days in advance. When they caught up with it they found that the protective fiberglass shroud had failed to detach, preventing the docking maneuver. Stafford radioed that the target looked "like an angry alligator."

On the second day Cernan crawled out of the hatch for a two-hour, nine-minute space walk—an extravehicular excursion in "a strange world," as he described it, 185 miles above the earth. He and Stafford were heard in radio conversation and, when Cernan's visor fogged, blanking his vision, the pilot ordered him to return to the cabin as the world below eavesdropped on the drama.

After traveling 1,200,000 miles on 44 orbits in 3 days, 25 minutes, Gemini 9 splashed down in the Atlantic 345 miles east of Florida only one-half mile from the planned target area and less than 4 miles from the U.S.S. "Wasp." For the first time TV cameras were near enough to scan the splashdown and recovery operations. Television from the deck of the ship relayed via Early Bird satellite showed the big candy-striped parachute as it dropped the capsule, the scene as it bobbed on the waves while frogmen attached the flotation collar, and then the hoist to the deck where the astronauts stepped out as the band played and the crew cheered.

"Communications were never so good throughout the flight," summed up an official at the control center. Radio enabled the flight surgeon to observe, "He's adapting very nicely to space walking," as he listened to Cernan's heartbeats as well as Stafford's. The Cernan beat was reported to have soared to 180 a minute at the peak of the maneuver and then settled to about 125. At the same time Stafford kept an eye on his companion and talked to him on the communications link in the 25-foot umbilical cord as millions of listeners around the world marveled at the fantastic achievement—a walk in space!

A LOT OF FUN

A three-day flight of Gemini 10 that began on July 17, 1966, was appraised as "probably the most productive and most important manned experiment in space to date." The triumphant spacemen were Commander John W. Young, U.S. Navy, and Major Michael Collins, U.S. Air Force. Launched from Cape Kennedy after a flawless countdown, they spent nearly 71 hours aloft and encircled the globe 43 times.

After a 103,000-mile chase the Gemini caught up with a 26-foot Agena launched 100 minutes ahead of it. Orbiting 185 miles out in space, the astronauts hitched on to it for nearly 39 hours and fired the Agena's rocket engine to blast deeper into space—476 miles, establishing a new altitude record for spacemen. It marked the first manned launching at orbital heights; it was the first use by a spacecraft of the fuel and propulsion system of another space vehicle to power its own flight. And to demonstrate man's ability to maneuver in space, a rendezvous was accomplished, but no link-up was attempted with another Agena in orbit about 247 miles high; it was a "dead bird" left over from the Gemini 8 mission in March.

When he was 245 miles above the earth, Major Collins stepped out of the open hatch on a 50-foot lifeline and "walked" over to the

Gemini 8 Agena, from the side of which he retrieved a micrometeorite detection plate of film, glass, and stainless steel designed to
register micrometeoric impacts. During his 38-minute trek he also
picked up an aluminum micrometeorite collection box mounted outside Gemini 10. But while closing the hatch the box slipped out of
hand and floated into space, as did one of his cameras. Various
scientific experiments were performed, and Collins on a second open-
hatch excursion took remarkable pictures of the earth as well as
spectrograms of stars on film sensitive to ultra-violet rays.

Throughout the journey ground controllers were in constant communication with the spacemen as the radio-television audience listened
and looked in. Commander Young was heard telling the spacewalker
Collins it was time to "come back into the house." The self-com-
posed astronauts were told from the world below them that they
were "doing a considerable job of maintaining radio silence"; replies
were generally terse and technical throughout the 1,300,000-mile
trip. While they were rated "the most untalkative of the astronauts,"
telemetry and radar kept "eyes" on them. The electronic scope in the
Control Center at Houston, Texas, enabled the TV audience as well as
the doctors to "see" their heartbeats calmly pulsing on the screen—
only 100 beats at blast-off, 110 to 130 for Collins during his space-
walk, and 80 beats per minute for both at retrofire in preparation for
re-entry, according to the TV commentator.

Excellent animated simulations enhanced by color intensified realism in following the flight on maps and through space as did the up-to-
the-minute commentary and interpretation from the Control Center.
The usual radio blackout injected about 5 minutes of suspense at re-
entry.

The aim was for splashdown about 500 miles east of Florida,
where the U.S.S. "Guadalcanal" and other ships and planes were at
the ready for recovery operations. The return to earth began above
the South Pacific when the braking rockets were fired 232 miles out
in space, 48 miles higher than at any previous re-entry. As in the case
of Gemini 9, the big parachute came down with the capsule within
view of the TV cameras on the deck of the carrier, just 3 miles from
the impact point and less than 5 miles from the ship. TV caught the
scene as a helicopter hoisted the spacemen from a rubber liferaft and
landed them on the "Guadalcanal's" flight deck. The dramatic panorama was relayed within range of a farflung TV audience via Early

Bird satellite, thus completing another triumph for communications in space as well as a step of considerable importance to the man-to-the-moon project. The ship's band blared, "It's a Big, Wide, Wonderful World," as the fourteenth manned space flight of the United States was completed as one of the most rewarding missions ever flown.

"We had a lot of fun," exclaimed Commander Young. "We were up over 400 miles and Columbus was right—the world is round!"

HISTORICAL SHARPSHOOTING

New records in electronic communications, navigation, and space mechanics were established by Commander Charles Conrad, Jr., and Lieutenant Commander Richard E. Gordon, Jr., both Navy pilots, who sailed into space on September 12, 1966, bound on a three-day, 44-revolution mission. In the first orbit, while 185 miles over Texas, the Gemini linked with an Agena rocket target launched three hours ahead of it. The swift performance of the feat was heralded as "historical sharpshooting in space—more than a dazzling display of technical virtuosity—a demonstration of the practicability of a maneuver." To which *The New York Times* added that space historians would have to invent new superlatives to describe the contributions of the Gemini 11 astronauts.

One of their experiments linked the Gemini and Agena by a 100-foot tether, and the Agena went into a slow spin. The centrifugal force created a trace of artificial gravity detected within the spacecraft. Commander Gordon took a 44-minute "walk" in space and later while standing in the open hatch he snapped pictures of the stars. In a historic maneuver the Agena's engine was used to thrust the docked Agena-Gemini to a new record altitude of 850 miles, the highest astronauts had ever gone. It was estimated that they could see 4,770 miles across the earth's surface from horizon to horizon, or an area five times that of the United States. As the spacemen described the fantastic view, flight controllers on earth were in constant communication.

When the time came to re-enter the atmosphere, the job of steering was turned over to a computer which guided the spacecraft to a safe splashdown in the Atlantic 700 miles east of Florida. Previous Gemini astronauts relied on computer guidance data but manually fired the braking and maneuvering rockets. The automatic aim of

Gemini 11 was so accurate that the capsule splashed on the sea only 2½ miles from the U.S.S. carrier "Guam." It was so close that the shipboard TV cameras caught a remarkable view of the 83-foot orange-and-white parachute as it dropped out of a blue sky with the capsule dangling beneath. Helicopters were seen hovering over as the spacemen were lifted from their raft and within minutes landed on the "Guam's" deck. The TV audience never had such a clear close-up view of such an event as the Early Bird satellite relayed the pictures to the nationwide networks; it was space communications at its best!

"This old world really looks good from the deck of this carrier," exclaimed Lieutenant Commander Gordon. "But I'll tell you something else again—it really looks great from 850 miles!"

LAST OF THE GEMINIS

Gemini 12 climbed into space over the Atlantic on November 11, 1966, with Captain James A. Lovell, Jr., U.S. Navy, and Major Edwin E. Aldrin, Jr., U.S. Air Force, aboard. Their first assignment was to rendezvous and dock with an Agena target vehicle launched several hours ahead of them. They photographed the earth and stars as well as a total eclipse in the Southern Hemisphere—the first such pictures ever taken above the earth's atmosphere. Major Aldrin on a two-hour "walk" in space—the longest then on record—successfully completed all of his extra-vehicular activities (EVA). He demonstrated that man can work effectively without excessive discomfort in the weightlessness of outer space; he showed that problems that beset previous astronauts, such as the visor becoming fogged, had been solved.

After a 94-hour, 33-minute flight in which they traveled 1.6 million miles on 59 circuits of the earth, they splashed down 700 miles east of Florida—right on time, right on target, only 3 miles from the recovery ship U.S.S. "Wasp." Television scanned the ocean scene while the Early Bird satellite relayed the pictures to the TV networks, enabling millions to witness the spectacular finale of the four-day mission, during which Lovell and Aldrin demonstrated the all-important role of communications, as did the astronauts before them.

When the 8,294-pound spacecraft splashed on the white-capped Atlantic it marked the end of the $1.35-billion Gemini project.. Brilliant records had been established in communications as well as in astronautics. A total of 40 Gemini days in space, and 17 million miles traveled in 609 revolutions of the earth, made history that encouraged scientists and engineers to believe they were on the right track in plans and equipment for landing men on the moon.[6]

Predicted *The New York Times* on November 14, 1966: "It becomes increasingly clear that men will land on the moon in the next two or three years; they will be on Mars and Venus well before the end of the century."

Tragedy at Cape Kennedy on January 27, 1967, put that prediction in doubt. A flash fire aboard the Apollo 1 spacecraft during a simulation test on the launch pad took the lives of Colonel Virgil I. Grissom, Lieutenant Colonel Edward H. White, 2nd, and Lieutenant Commander Roger B. Chaffee.

COUNTDOWNS STOPPED BY TRAGEDIES

The timetable to the moon was slowed by many months beyond expectations. Nearly two years would pass before another manned Apollo redesigned and vastly improved would go on an 11-day earth-orbital flight to test the way for a circumlunar journey in 1969.

While all phases of the Apollo project were under review with delay and drastic changes in prospect, the Russians again went into space. Soyuz 1 (signifying union), piloted by Colonel Vladimir M. Komarov, who commanded Voskhod 1 in 1964, was rocketed into orbit on April 23, 1967—the first Soviet manned flight in more than two years. Traveling 18,641 miles an hour, the spacecraft circled the earth every hour and 28 minutes at a maximum altitude of 137 miles, minimum 124 miles.

On the seventeenth orbit Komarov radioed he was having difficulty controlling the craft, and was preparing to return to earth. After re-entry the main parachute lines entangled; the chute failed to open causing the spacecraft to plummet 4.3 miles, killing the 40-year-old cosmonaut.[7]

This tragedy and the one at the Cape led to revised planning by both the United States and Russia. Countdowns stopped.

[6] Project Mercury involving six one-man flights was estimated by NASA to have cost $400.6 million; Project Gemini, ten two-man flights $1.3 billion.

[7] Colonel Yuri A. Gagarin, the world's first astronaut, was killed in a plane crash on March 27, 1968.

VII   Trail-Blazing the Lunar Age

> It is . . . . time for this nation to take a
> clearly leading role in space achievement,
> which in many ways may hold the key
> to our future on earth.
>
> —*John F. Kennedy*

ALL PHASES OF COMMUNICATIONS SHARED IN THE ASTRONAUTS'
achievements. The global tracking and communications network
played such a vital role that attention was focused upon the necessity
for development of a flawless communications system between the
earth and moon, lest lunar voyagers be lost in space.

When and what next? Project Mercury having terminated, the
National Aeronautics and Space Administration announced that
plans called for the Gemini project—a manned flight in a two-man
capsule—and later the Apollo lunar-landing project—a man on the
moon. It might cost $24 billion in over a decade.

Recalling that Cooper's flight had marked the thirty-sixth anniversary of Lindbergh's solo hop to Paris, President Kennedy said,[1]
"Both flights were equally hazardous. Both were equally daring. I
know that a good many people say, 'Why go to the moon?' just as
many people said to Lindbergh, 'Why go to Paris?' Lindbergh said,
'It's not so much a matter of logic as it is a feeling.' " And the
President added, "I think by the end of the sixties we will see a man
on the moon—an American."

[1] Greeting Major Cooper at the White House, May 21, 1963.

When, in 1865, Jules Verne wrote *From the Earth to the Moon,* he said it was logical, in view of "the audacious ingenuity" of Americans, to expect them to reach the moon.

SOME KEY QUESTIONS

"From the scientific standpoint, exploration of the moon is of great importance," said James E. Webb, administrator, National Aeronautics and Space Administration.[2] "The moon may hold the answers to some of the key questions in science. How was the solar system created? How did it develop and change? Where did life originate? The moon is devoid of atmosphere in the terrestrial sense. Having neither winds nor rains, its surface is almost changeless. Thus the moon offers scientists a chance to study the very early matter of the solar system in practically the form in which it existed billions of years ago.

"For the sake of this country's prestige, it is essential that other nations believe that we have the capability and determination to carry out whatever we declare seriously that we intend to do. Space achievements have become today's symbols of tomorrow's scientific and technical supremacy in the minds of millions of people. The manner in which knowledge emerging from space science and technology is put to use will become more and more persuasive in the battle for men's minds. . . . In its very essence, science is international. . . . In marshalling and developing scientific and technical resources required for manned lunar exploration, we shall be creating a technology that is certain to radiate great and diversified benefits to almost every area of material and intellectual activity."

Already, Mr. Webb said, the national investment in space exploration has produced new materials, metals, alloys, fabrics, and compounds which have gone into commercial production. From work in space vacuum and extreme temperatures have come new durable, unbreakable plastics and new types of glass that will have a wide variety of uses. Advances also are being made in the development of

[2] *The New York Times,* October 8, 1961. (Webb, appointed administrator of NASA by President Kennedy in February 1961, resigned in September 1968. Dr. Thomas O. Paine was appointed administrator by President Nixon on March 5, 1969).

thermoelectric materials, for example, a germanium-silicon alloy with which heat can be converted directly to electricity. Not only have thousands of new products and new materials been developed but communication techniques, equipment, and systems are being revolutionized for the utilization and exploration of space.

"Medical scientists in the space effort," continued Mr. Webb, "have devised minute sensors to gauge an astronaut's physical responses and to measure his heartbeat, brain waves, blood pressure and breathing rate. These same devices could be attached to hospital patients so that their conditions could be recorded continuously and automatically at the desk of the head nurse.

"More than 3,200 space-related products have been developed in the United States. They come from 5,000 companies and research organizations engaged in missile and space work. From this new industry are coming new opportunities and new jobs. . . . Space exploration in general and the manned lunar program in particular, offer the United States the chance for unparalleled material progress and for maintaining our position in the world. . . . I am convinced that most Americans realize that we could not be bystanders to any key phase of this tremendously challenging adventure, with all its implications for the future of mankind."

A CREATIVE CHALLENGE

The moon brightened as a creative challenge that extends across science and industry, communications, and military operations. Americans were enchanted by the lunar prospects. By pioneering they achieved leadership, often, however, at a much higher cost than if they had patiently let nature take its course. Color TV is a good example; Americans had it first.

Space exploration to gain more knowledge of the universe could be—as a leading educator put it—"one of the great scientific achievements or enterprises of all time; its impact on the world and mankind is simply beyond calculation."

Inspired by that outlook, a new unity has been established among statesmen, physicists, chemists, metallurgists, astronomers, engineers, and scientists in many fields from aviation to electronics and communications. All work together, whereas in former times each group went its own way to seek knowledge and to make discoveries. Space brings them all together, functioning as a team on a common ground in communications, propulsion, weather forecasting, and in the development of spacecraft. The technology of space is expanding so rapidly that it has become the glamorous branch of science, industry, and education.

WHY SPEND $24 BILLION?

To those who ask, "Why spend $24 billion to go to the moon?" Representative George P. Miller, chairman of the House Committee on Science and Astronautics, supplied an answer in a speech before the American Astronautical Society in Washington on January 16, 1962: "The by-products of this effort will have economic, social, and defense values far in excess of the original cost. We know even at this early stage of development and research that space exploration is of such importance to man's total knowledge that it will benefit and alter the course of his existence in ways no more foreseeable today than those which resulted from the invention of the wheel."

The big question directed to Marconi at the turn of the century was, "Is it worth $200,000 to send a wireless signal across the Atlantic?" Of course, that kind of money was not so linked with the national economy and taxes as in the case of the billions required to put a man on the moon, which involves thousands of companies.[3]

Sixty-one years later—in 1962—the United States Government put more money into research and development than it spent in the entire interval from the American Revolution through and including World War II, according to Dr. Jerome B. Wiesner, director of the Office of Science and Technology and special adviser on science to the President, in a report to Congress in July 1962.

Despite Presidential enthusiasm on the race to the moon and the vast expenditures required, Congressmen as well as others questioned

[3] In 1966, NASA had 35,700 employees and indirectly 400,000, through industrial and university contractors; in 1969 staff members totaled 31,700 and indirect employees approximately 185,000.

the necessity of being first on the moon.[4] By 1966, the number of critical voices in the congressional debate on the budget for interplanetary projects increased. Fundamentally it was questioned whether the United States should have become involved in a moon race in the first place. Invidious comparisons were drawn about the dividends from money spent in space and those that might be realized if those funds were used to meet urgent needs of "inner space." It was contended that priority should be given to the war on poverty, crime, and pollution; also to health, education, urban renewal, and modernization of transportation.

Space budget critics declared "enough is enough." Their impatience became more vociferous, agitated by what was termed extravagance in "moonship follies." They asked, "Is a bag of lunar rocks and a box of moon dust worth $24 billion?" Scientists said "yes"; they appraised the rocks as priceless, and precious.

The idea of a marathon to Mars and possibly to Venus touched off intensified opposition to the space budget. A voyage to the ruddy planet would be mighty expensive, with estimates running as high as $100 billion. It would be an endurance test not only for the budget but for man himself; he might be a year enroute on a long curving path, then spend weeks exploring before beginning a year-long trip back to earth. The distance between Mars and the earth varies over a wide range because of the large eccentric Martian orbit that brings the planet nearest the earth—about 35 million miles—once every 17 years.

Critics declared it would be deplorable—inexcusable—for the dash to the moon to be succeeded by a race to land a man on Mars, already photographed by unmanned spacecraft as a desolate, lifeless orb. It became more and more apparent that the magnitude of space exploration called for a general undertaking—unified effort by all nations, not only individual countries, if for no other reason than to avoid duplication, reduce expenditures, and share technological know-how.

Perhaps the day of collaboration would come. But until then international rivalry, national prestige, and politics, as well as ideological

[4] NASA's budget increased from $117 million in fiscal 1958 to a high of $5.9 billion in 1966; $5.3 billion 1967; $4.01 billion 1968; $3.76 billion 1969; $3.7 billion 1970.

glory, seem destined to persist as incentives to earmark huge sums for spacemanship and to keep the crucible of space at a boiling point for a long time to come.

ENTRY OF SATURN AND APOLLO

When Saturn 5, a moon-rocket proclaimed as the mightiest of space machines, hurtled into the skies on November 9, 1967, hope soared anew that American astronauts would walk on the moon before the end of the decade. The 6.2-million-pound, 363-foot shaft, a 36-story giant with a three-stage rocket capable of generating 160 million horsepower, or 7.5 million pounds of thrust at lift-off, was the largest ever flown.

The 8½-hour maiden flight won Saturn acclaim as "the most complicated single integrated machine men have ever built and successfully operated." So thunderous were the sound waves emitted by the spectacular blast-off from Cape Kennedy that sensitive microbarographs recorded them in New York. A feature of the 140-ton payload—the heaviest ever orbited—was an Apollo spacecraft, a 12-foot-long cone designated Apollo 4 and known as a "command module," like the one foreseen carrying astronauts on a lunar voyage.

The capsule was thrust up 11,234 miles, from which point it plunged toward the earth at 25,000 miles an hour, about 7,000 miles faster than the re-entry speeds of the Gemini spacecrafts. Communicators tuned for evidence that the heat shield had withstood the terrific re-entry temperatures up to 5,100 degrees above zero Fahrenheit encountered when the craft struck the atmosphere at an altitude of 78 miles. Then, 18 minutes after re-entry, a parachute dropped the capsule from 23,000 feet to the splashdown about 700 miles northwest of Hawaii, where the U.S.S. "Bennington" was nearby for recovery.

President Johnson evaluated the Saturn performance as proof "we can launch and bring back safely to earth the spaceship that will take men to the moon."

Man could boast that he was no longer limited in travel to land, water, or air, or by the gravity of the earth. "We can overcome those limitations," declared James E. Webb, NASA administrator, "and move out any place we really want to go."

TEST OF A MOONSHIP

As evidence that he might be right, the first flight model of an unmanned Apollo moonship—a 16-ton lunar module (LM) was blasted into space by a Saturn rocket on January 22, 1968. The bug-shaped vehicle was both a two-stage rocket and a two-man space-craft. With no man aboard, a computer served as an electronic guidance system to provide the on-and-off signals and throttling cues to the descent engines as well as other controls.

On a lunar landing mission three astronauts would be rocketed toward the moon in an Apollo command capsule. When in lunar orbit two men would crawl into the attached lunar module, separate it from the orbiting Apollo, or mother ship, and descend to the moon. With the mission accomplished they would climb back into the module and return to the hovering Apollo for the journey back to earth. The first unmanned, earth-orbiting, eight-hour flight rated a success—all essential objectives accomplished—the pioneering LM continued in orbit for about three weeks before disintegrating as it re-entered the earth's atmosphere.

A Saturn 5 designated as Apollo 6 was not as successful. Lofted on April 5, 1968, to an altitude of 13,800 miles, it could have been the final unmanned test, but engine flaws marred the planned ten-hour flight. It failed to gain the proper orbit and splashed in the Pacific to be picked up by a Navy helicopter.

APOLLO 7'S ODYSSEY

"All go" was declared for Apollo 7 on October 11, 1968. Rocketed by a Saturn 1-B booster with a 1.6-million-pound thrust, the 6-ton cone-shaped capsule carried a three-man crew: Captain Walter M. Schirra, Jr., U.S. Navy, Major Donn F. Eisele, U.S. Air Force, and R. Walter Cunningham. A perfect lift-off sent them racing through space on an 11-day earth-orbit that would test the way for a circumlunar flight. TV viewers watched pictures from an Air Force camera IGOR (Intercept Ground Optical Recorder) that followed the rocket for more than 100 miles.

This was a radically new spacecraft redesigned and vastly improved at a multi-million-dollar cost—estimated at $400 million— since the disastrous fire on the launch pad in 1967. Now the outer space suits were made of Fiberglas Beta yarn that withstands temperatures 400 degrees Fahrenheit to 1,200 degrees F. In-flight garments were of the heat-resistant fabric Teflon fiber designed to remain stable and offer flex-abrasion resistance under extreme chemical and temperature conditions.

Featured among many other modifications was a hatch that could be opened in 7 seconds instead of 90 should trouble develop on the pad prior to blast-off, and stainless steel jacketed wires. Design specifications ranging from the massive to microminiature called for safety and fireproofing.

While on the pad a gas mixture of 60 per cent oxygen and 40 per cent nitrogen was substituted for the 100 per cent oxygen environment in the capsule during flight. It was reported that at least 27 different fireproof and fire-resistant materials were substituted for more easily ignited materials aboard the earlier Apollo. The spacecraft's liquid oxygen and liquid hydrogen tanks were so leak-proof that it was estimated an auto tire, leaking at an equivalent rate, would not go flat for 32 million years. And the tanks were so well insulated that ice cubes inside would take more than eight years to melt, according to the statistical analogies.

Apollo 7 also introduced numerous advances in communications as demonstrated from take-off to landing. The blast-off seen on television throughout the United States was relayed via satellite to Canada, Europe, and Japan. A 4½-pound RCA mini-TV camera operating with a 6-watt transmitter facilitated live telecasts from a United States spacecraft for the first time.

While Apollo streaked across the southern region of the United States at 17,500 miles an hour, clear pictures of the astronauts in the cabin and views of the earth were beamed to earth. The spacemen were seen floating around in weightlessness while a pen and other objects drifted by the lens. They held up signs, one of which read, "Keep those cards and letters coming in, folks." Another was lettered, "Hello from the lovely Apollo Room, high atop everything."

As the "Apollo Room," up from 103 to 130 miles, moved out over the Atlantic 800 miles east of Cape Kennedy's tracking antenna, "the first TV show from a U.S. spacecraft" began to fade and disappeared until another orbit over Texas and Florida. Significant military, economic, and geological values were foreseen for TV photography from outer space.

During 163 orbits—one every 90 minutes—the flight path ranged from about 100 miles to 266 miles high. While the astronauts traveled 4½ million miles, they performed numerous experiments, including a successful rendezvous with their discarded Saturn 4B booster.

After a flight of 260 hours, 8 minutes, the spacecraft came down on October 22 some 325 miles south of Bermuda with re-entry controlled automatically by an on-board computerized guidance system, which led to splashdown about 8 miles from the U.S.S. "Essex." Low-hanging rain clouds prevented TV cameras on board the carrier from seeing the parachute or the descent on the sea. Three helicopters were in the air looking for the Apollo hidden in the early-morning haze and in radio silence caused by the fact that the capsule landed upside down, submerging the antenna. Suspense ended when the craft was righted by balloons inflated outside the capsule and communication resumed. The helicopters reported reception of the homing signals and sighting of the spacecraft. Radio voices soon told that the flotation collar was in place, the hatch opened, and the bearded spacemen climbed into a rubber raft from which they were lifted to the rescuing copter. Within a few minutes they were ferried through the rain squalls to the deck of the "Essex," where TV scanned the end of "a perfect mission—a very major step toward the manned lunar landing."

THE RACE RESUMED

Four days after Apollo 7 dropped back to earth, Colonel Georgi T. Beregovoi took to space on October 26 in Soyuz 3. Evidence prevailed that the race to the moon had resumed full-tilt.

Launched ahead of Soyuz 3 was unmanned Soyuz 2 for automatic and manual control maneuvers of spacecraft while they rendezvoused at nearly equal speeds about 88 miles up in circular orbits. After some 40 spins around the globe, Soyuz 2 was returned to earth while its counterpart continued in orbit with Beregovoi using TV that showed him in the cabin as well as views of the earth. Having conducted "scientific and technical research" on the fourth day after 64 orbits, the astronaut brought his ship to a soft landing in the steppes of Kazakhstan, and reported "everything had gone according to plan."

Lieutenant Colonel Vladimir A. Shatalov piloted Soyuz 4 into earth orbit in January 14, 1969. Blast-off at the Baikonur Space Center in Kazakhstan was telecast, as were scenes within the cabin as Shatalov encircled the globe every 88 minutes.

The next day Soyuz 5 with three cosmonauts aboard: Lieutenant Colonel Boris V. Volynov, Aleksei S. Yeliseyev, and Lieutenant Colonel Yevgeny K. Khrunov went into space and linked up in a docking for more than four hours. Two of the Soyuz 5 crew "walked" in space and entered Soyuz 4 for an hour visit while the vehicles orbited between 155 and 130 miles above the earth. Television cameras installed inside and outside the two spacecraft enabled Russians to watch the link-up, the transfer of the spacemen, and the undocking. The achievement was heralded as "the world's first experimental space station" destined to lead to the relief of crews aboard permanent space stations, or the rescue of astronauts in distress.

After 49 earth orbits on a three-day flight, Soyuz 5 landed in northwest Kazakhstan. Soyuz 4 continued in space for another day before returning to earth, having made 45 orbits in 71 hours and 14 minutes. A 7-cosmonaut mission Soyuz 6, 7, and 8 was launched in mid-October, 1969. The troika was reported to have conducted scientific experiments and maneuvers in orbits about 140 miles above the earth. Each vehicle landed after the 80th revolution and five days in space.

APOLLO 8'S FOURTH SEAT

Communications, flight, science, and exploration were given extended dimensions when Apollo 8 and its three-man crew as Columbuses of space blazed a trail to the moon and sailed around it ten times in 20 hours and 14 minutes. Communications was a lifeline unreeled through the vastness of the solar system. In marked contrast, the caravels of Christopher Columbus on the transatlantic voyage to "the new world" in 1492 were lost from sight and in silence for seven months. Not so with Apollo. These heroic explorers conversed with earth; they were seen in the moonship's cabin; they televised the sunlit western hemisphere so that millions of earthlings could see "themselves" as the Man in the Moon had looked at them for centuries.

Blasted skyward by a Saturn 5 rocket at Cape Kennedy on December 21, 1968, Apollo carried a trio of spacemen to be remembered in history as well as in the annals of communications: Colonel Frank Borman, U.S. Air Force, Captain James A. Lovell, U.S. Navy, and Major William A. Anders, U.S. Air Force. They were expert communicators as well as superb astronauts. Rocketing from the earthly abode at an initial speed of 24,200 miles an hour, they bravely embarked on a hazardous journey never before attempted by man.

When 139,000 miles out on the trail, they turned on the TV and telecast pictures of them in the cabin and a view of the earth that looked like a bright ball of light in the heavens. They described it as a blue sphere veiled in clouds.

A televiewer, as if looking out the window of the spaceship, wondered how 3.5 billion people—and double that by the turn of the century—could be living on such a puny sphere floating in space among the stars. There was not the slightest evidence of man's handiworks—skyscrapers, bridges, super-highways, industries, or millions of automobiles rushing hither and thither on the receding globe. Where were the Seven Wonders of the World—Niagara Falls, the Grand Canyon, the pyramids, etc.? No inkling of them! The wonder of the world at the moment was Apollo 8, its remarkable pilots and their use of radio-television. The earth so far away looked more like a grapefruit than the hectic orb it is. At times on solar TV the world appeared like a small ball as the picture bounced around the screen. One wondered how such a blob could contain war, hostility, racial unrest, violence, politics, a wide variety of nationalistic interests, along with a multiplicity of other things. With the globe shrunk to such proportions earthly problems and bickering seemed petty and unimportant.

As Apollo sped further away into the moon's sphere of gravitational influence, only the earth's continents and seas were recognizable, but soon they too vanished in the distance. The pendent globe took on the aspect of an insignificant microcosm that inhabitants of other worlds—if they exist—might well consider to be an uninhabited dot of light in the enormity of the universe. Said Lovell, "What I keep imagining is if I am some lone traveler from another planet what would I think about the earth. Whether I think it would be inhabited or not."

Television, "the fourth seat" in the spacecraft, dramatically demonstrated man's technical virtuosity—a revolution in modern communications. Both radio and television were making history that paled to insignificance the old communication records established on earth since Marconi's transatlantic signal, heralded at the time as the first great triumph of the twentieth century.

The 4.5-pound TV camera, identical with that of Apollo 7, was equipped with a 160-degree wide-angle lens for on-board monitoring and a 100-millimeter lens for scanning outside the spaceship. Miniaturized integrated circuits facilitated design of the camera 30 times lighter and 85 times smaller than a standard TV broadcast camera. And the Apollo camera required only 6 watts for operation compared with 500 watts for a studio camera. The video signals picked up at Goldstone, California, and at Madrid, Spain, were processed and relayed via cable and microwave circuits to the NASA Manned Spacecraft Center, Houston, Texas, for transfer to the TV networks.

Ground-station scan converters made the Apollo signals compatible with home TV receivers. Communication satellites hovering high above the Atlantic and Pacific relayed the pictures to Europe and Asia. Unparalleled electronic magic was performed between the spaceship and the home sets that tuned in the live broadcasts from lunar orbit as people everywhere marveled at the pioneer explorers "voyaging in a realm where, so far as men know, life of any kind has never been seen."

The second telecast from 176,000 miles featured sensational pictures of the earth. Lovell said that he could distinguish through binoculars the mouth of the Mississippi, the Gulf of Mexico, all of the South Pole, but not much of the North, hidden in clouds and darkness. He talked back and forth with the Control Center as if on a local telephone line. While millions of spectators eavesdropped, he described the oceans as royal blue, the land areas as brownish, clouds as bright white, and reflection from the earth as much brighter than the moon.

As the spacecraft orbited the moon, the third telecast presented the lunarscape scanned through the window from an altitude of 70 miles. Commenting on the countless craters, the astronauts conducted a guided tour of the moon. Anders said the moon looks "like whitish gray beach sand with lots of footprints." Much happened in split seconds between the moonship and armchair explorers in living rooms throughout the world. The camera's signal flashed to the earth station at Robledo, Spain, hence to a Spanish-owned terminal in Bulitrago, Spain, from where it leaped up to a communications satellite over the Atlantic for relay to Andover, Maine. From "the big ear" at that site the signal was fed to Western Union International in New York for routing to Houston and the TV networks.

As Americans saw the pictures and looked up at the crescent moon on Christmas Eve, the Apollo triumph seemed unbelievable. Children all around the "oasis" were watching for Santa Claus and his reindeer to come racing out of the sky, while giant telescopes at Denver and other astronomical centers focused on the moon in hope of a reflected speck-like glimpse of Apollo. It remained secluded in the distance, however, as the astronauts signed off for the evening, taking turns in reading passages from the Book of Genesis about the creation of the earth.

A phenomenal lunar tour highlighted the final encirclement. Amid the spotting of giant craters, mountains, and "seas," the astronauts radioed their impressions a few hours before firing the rocket that would thrust them out of lunar orbit and begin the 57-hour return trip to a miniature earth described by Lovell as "a grand oasis in the great vastness of space."

Borman beheld the moon "a vast, lonely, forbidding-type expanse of nothing. It certainly does not appear to be a very inviting place to live or work."

"You can see by the numerous craters," added Anders, "that this planet has been bombarded through the eons with numerous small asteroids and meteoroids pockmarking the surface, every square inch of it."

Rewarded with magnificent color movies and hundreds of photographs as spectacular proof of what they had seen, including the hidden face of the moon, the spacemen turned toward home. The moon was left behind as Apollo crossed the celestial divide between the lunar and the earth gravitational grip. It "coasted" back toward the world at increased speed that would reach 24,530 miles an hour when the spaceship lunged into the earth's atmosphere.

When 191,750 miles away, the astronauts staged their fifth telecast—a tour of their 33-foot moonship "home," as they called it. No expedition ever had such public participation. Television seemed to put every viewer into the capsule. And on Christmas Day the spacemen were TV visitors at the fireside in millions and millions of homes.

When about 111,500 miles homeward bound, the TV camera scanned the earth being approached at about 4,100 miles an hour. The fantastic 500,000-mile voyage of 146 hours, 59 minutes' duration heralded as "a superlative feat of modern history" was nearing an end. Borman, Lovell, and Anders had ventured farther than any man had ever dared and escaped the earth's gravitational pull for the first time. Their altitude record was registered at 234,672 miles.

The next day—the sixth of the epic $310-million flight—in predawn darkness about 1,000 miles southwest of Hawaii, Apollo made a flawless, pinpoint splashdown within 4 miles of the U.S.S. "Yorktown." Within the hour Navy frogmen dropped from a helicopter to attach the flotation collar around the capsule charred by the 5,000-degree heat in the searing descent at re-entry. While the spacecraft bobbed on the waves awaiting hoist to the deck of the aircraft carrier, television favored by the light of sunrise showed the first circumnavigators of the moon stepping out of a helicopter on the ship's deck as cameras scanned the historic scene for transmission around the world.

Shakespearean students recalled that some 350 years ago Hotspur in *King Henry IV* remarked:

> "By heaven, methinks it were an easy leap
> To pluck bright honour from the pale-faced moon . . ."

It was not as easy as Hotspur thought. Nevertheless, precision navigation, electronic communications, countless achievements in aerodynamics, astronautics, and engineering, so magnificently maneuvered by the Apollo astronauts, enabled them to pluck "bright honour" and immortality as they photographed and telecast the pale face during their ten circumlunar voyages.

The moon, however, having been seen close up as a barren, desolate, forbidding, pockmarked wasteland, might well have justified televiewers to wonder why risk other lives to land a man on such an awesomely rugged terrain at a cost of billions of dollars that could be used so beneficially to advance living on earth?

Scientists took a different view. They looked to the future. If man could make a reconnaissance of the moon, all sorts of prospects and opportunities were foreseen. It could become a vital base for celestial exploration—a launch pad for expeditions and travel. It might become a scientific laboratory unhampered by earthly impedimenta. It could serve as a military outpost and as an astronomical observatory uninhibited by the earth's atmosphere and distance. It might well qualify as an interplanetary communications center.

"Space is not a gambit," President Johnson once remarked. "It is not a gimmick. Where the moon is a major goal today, it will be tomorrow a mere whistle stop for the space traveler."

Exploration of space is such a challenge to American technology and American industry as has never happened except under the stress of war, Borman declared at the 1969 conclave of the American Publishers Association. "That technology and that capability," he said, "is transferable into every phase of our life." Furthermore, along with technology, the space program had produced a remarkably capable group of young engineers. For example, the average age of those who computed the Apollo 8's trajectory was only 23.

Envisaging news of the future, Borman told the news executives that space plans called for nine Apollo lunar module landings at different sites on the moon, which he pointed out is about the size of Africa. And no one would claim to have explored that continent by landing only once in the Sahara. Next he foresaw large multimanned scientific space stations like the Antarctica project. Then "in the not too distant future" should come a trip to Mars and to other solar systems.

Perhaps the most important reason for reaching out to other horizons, he said, might be that "people the world over will start to think of themselves not merely as Russians, Chinese, or Americans but perhaps as earthmen." He had wondered when he looked over the lunar horizon at the earth as a fragile piece of granite "in the midst of black nothing," how human beings could produce the mechanical miracle of Apollo and "yet in all recorded history not have been able in some way to live together in peace and harmony."

Congress in a joint session on January 9, 1969, saluted the Apollo 8 crew as "history's boldest explorers" and cited them for "significantly advancing the nation's capabilities in space." In response, Borman said, "I'm convinced that it is no longer whether we'll do these things, it's how long it will take and how much we will spend. Exploration is really the essence of the human spirit, and I hope we never forget that."

APOLLO 9 AND THE SPIDER

Coupled to a lunar module (LM) Apollo 9 was thrust into the heavens by a Saturn 5 rocket on March 3, 1969, to conduct a ten-day flight test of all equipment and to rehearse procedures for a moon landing. Color television's electronic artistry dramatically painted the awesome orange-red flames that enshrouded the launch pad as the 6.5-million-pound rocket thundered upward, then leaned eastward out over the Atlantic on a mission described as the most difficult, most complex space experiment man had yet attempted. The primary purpose was to assay the spidery-looking, 16-ton lunar module and practice flying it in rendezvous maneuvers with the command ship. Success in proving the LM spaceworthy was vitally important; future astronauts returning from the moon would have to simulate the tactics in order to get back to earth.

Colonel James A. McDivitt and Colonel David R. Scott, both of the U.S. Air Force, were accompanied by Russell L. Schweickart, a civilian. Quickly they went to work. Three hours after a perfect launch, when traveling 17,400 miles an hour, 118 miles over California, the command ship pulled away, leaving the lunar module still attached to the third stage of the Saturn. Protective "petals" flew off fully exposing the LM, toward which Apollo made a U-turn for a nose-to-nose link-up. The rocket stage stayed behind; on command from earth, rockets were fired to send it out to orbit the sun. With the command ship and the module now flying in tandem, Colonel McDivitt, a veteran space communicator, radioed, "Everything came off just right. We're hard docked."

On the second day the propulsion engines fired to toss the locked-on spacecraft into higher orbits ranging up to 312 miles for rigorous experiments and practice flying of the two attached vehicles. All systems having been proved A-OK, McDivitt and Schweickart on the third day crawled through the connecting tunnel into the moon-lander and fired one of its main rockets. They tested all phases of the LM's intricate communications system designed to beam everything from the astronauts' pulse rates to live TV, which showed them in the cabin as they explained various instruments, ate lunch, and took pictures out the window. Two radio systems on board the module featured VHF (very high frequency) transreceivers for contact with the command ship and an S-band system for LM earth links.

On the fourth day Schweickart opened the hatch and stepped out on the LM's "front porch" for 38 minutes, to behold "what a view!" from 147 miles up. He was a self-contained human spacecraft—the first American to leave an orbiting spaceship depending solely on a back pack for oxygen, air-conditioning and communications rather than on "umbilical hoses" or support lines leading from the craft. His $2-million suit was the same as astronauts would wear when walking on the moon. Conversation between Schweickart and McDivitt was heard loud and clear on earth while the "walker" stood on the porch-like platform testing the space suit and taking pictures as well as thermal samples. Television simulations geared to the live radio voices were classic demonstrations of the astronauts in performance. Televiewers had the illusion of being right there with them.

The fifth day's maneuver was to separate the two craft and fly the LM about 100 miles away; then, relying primarily on radar, find the way back to rejoin the command ship. Radio kept McDivitt and Schewickart in communication with Scott, who was piloting the mother craft, and with the ground controllers. The docking feat was heralded as spectacular two-vehicle navigation—a stunning triumph for American technology.

Radio listeners marveled at the achievement revealed by the space talk as the two craft approached for the delicate locking maneuver, depicted realistically by excellent simulations on TV. As the LM hove in sight Scott was heard to say, "You're the biggest, friendliest, funniest-looking spider I've ever seen."

The LM's octagonal fuselage, the antenna array, radar gear, and spindly legs gave it the appearance of an insect, thus the nickname "Spider." The incredibly intricate spacecraft was conceived by Dr. John C. Houbolt, a NASA engineer who became senior vice president of the Aeronautical Research Association, Princeton, New Jersey.

His blueprint was followed by the Grumman Aircraft Engineering Corporation in construction of the lunar lander under the supervision of its chief designer Thomas J. Kelley.

The cone-shaped commandship was known as "Gumdrop," a code name adopted by the astronauts when they saw it arrive from the factory swaddled in a blue plastic covering that reminded them of a gumdrop.

The $340-million mission's crucial objective of proving the LM spaceworthy having been proclaimed a historic success, the expendable and expensive Spider representing an estimated $100 million worth of space hardware was cast adrift to go into a higher orbit from which eventually it would drop and burn up in the providential atmosphere that shields the earth from the perils of such wanderers gravitating toward the globe.

The flight plan for the final five days specified engine firing that would put the Apollo in various orbits ranging from 109 to 242 miles high, including the final orbital adjustment before re-entry. The astronauts also began landmarking assignments, and infrared photography to aid in the study of crop conditions and natural resources as scanned from space. From blast-off to splashdown an ingenious electronic guidance and navigation system coordinated and controlled by an onboard computer did the driving and automatically mastered firing of engines to maneuver the craft.

During the fiery re-entry radio communication was disrupted. Several minutes passed before the suspense was punctuated by reports that radar had spotted the descending spacecraft. As it continued to plunge earthward communication was restored and the astronauts' voices came through again. Triple orange-and-white parachutes opened and soon the TV cameras onboard the recovery ship U.S.S. "Guadalcanal" focused on the huge umbrellas swinging the capsule in the bright blue sky 360 miles northwest of Puerto Rico.

The pinpoint splashdown some 3 miles away from the ship afforded an international TV audience a grandstand view of the heat-scarred spacecraft bobbing on the sea. Frogmen were seen attaching the flotation collar and inflating yellow and orange life rafts for the spacemen to jump aboard as they climbed out the hatch. A helicopter lowered the basket-like "bird cage," and one by one the astronauts were hoisted aboard; a short ride put them on the deck of the carrier for a red-carpet welcome. Their historic 3.7-million-mile journey had ended.

Throughout the entire recovery operation color television was spectacularly perfect, especially as the ship moved within 100 yards of the capsule; the close-up views were incredible, vividly realistic—video communication at its best! Television's renowned characteristics of immediacy and intimacy of perspective were never better illustrated. So clear were the pictures in breathtaking detail that for anyone whether looking in on Long Island or Hawaii, London or Toyko, it seemed unbelievable that this news event was happening so many miles away; it was like looking out a window at home.

Part of the success was attributed to a breakthrough in satellite communications technology. Erected on the deck of the "Guadalcanal" was a new portable "ground" station built by General Electric for Western Union International. A 15-foot dish antenna beamed the TV scene to the ATS-3 satellite, which relayed to the NASA ground station at Rosman, North Carolina. From that terminal the graphic scenes were fed to the nationwide networks and via satellite overseas.

FIRST COLOR TV FROM SPACE

It remained for Apollo 10 "to tie together all the knots and sort out all the unknowns" as its commander Air Force Colonel Thomas P. Stafford explained the mission to which he and Navy Commanders Eugene A. Cernan and John W. Young were assigned. Their eight-day journey—the last rehearsal before a lunar landing—began on May 18, 1969.

An epochal record in communications was established when the first live color TV from 22,781 miles showed the earth smaller than a basketball suspended against the black background of outer space. In color too the command module was seen edging up to dock with the lunar module. And inside the spacecraft a 12-pound, $75,000 camera built by Westinghouse to operate on 20 watts scanned the astronauts in the cone-shaped cabin and peered out upon the firmament and the void in which it floats.

When 115,000 miles away, almost halfway to the moon, the TV camera again went into action. The tiny earth, a bluish, cloud-covered sphere, bounced around on the screen. Seeing it as such a microscopic dot in the universe—packed with everything that characterizes life on the globe—revealed the wonder of creation, and indeed mastery in micro-miniaturization ad infinitum! "Looking down from here," said Stafford, "you'd never know anybody inhabited the place." Every armchair "astronaut" must have agreed. And there were an estimated 80 million of them.

To see the earth shrunk to such infinitesimal proportions and to think of it as its inhabitants know it, projected against the fact that the creation was completed in six days, comprised a sermon for mankind to hear and behold.

When 210,000 miles from home the receding globe appeared smaller and smaller, described by Stafford as slightly smaller than a tennis ball and a little bigger than a golf ball. Earthlings shared the incredible view on color TV that came in through the 210-foot "dish" antenna at the Goldstone Tracking Station in California.

As the spacecraft went into 62 hours of lunar orbits ranging from 196 to 69 miles above the moon, it passed behind the sphere and radio contact ended for 27 anticipatory minutes. Suspense became more intense with the countdown of the minutes on the TV screen. Millions were anxiously listening for news whether the main rocket engine had a good burn that put the Apollo into lunar orbit for 31 two-hour trips of circumnavigation. When the countdown hit zero exactly as scheduled, the radio signals were reacquired as the spacecraft moved out from behind the moon and Stafford exclaimed, "You can tell the world we have arrived!"

Within ten minutes after reappearing, television presented a spectacular panorama of maria (lowlands or "seas"), craters, rilles, deep valleys, and huge boulders. "Boy, this is really a rugged planet," declared Young.

Voice commentary by the spacemen and the clarity of conversation with the Control Center were extraordinary, indeed communication true to the old phrase "out of this world." They demonstrated weightlessness as they nudged various objects to float around the cabin; they switched on pretaped melodies, "Up, Up, and Away" and "Fly Me to the Moon"—the first tunes to reach the earth from that far away! All the dreams of Jules Verne and Marconi were becoming reality.

"You know, you can study all your life and never finish studying," said Stafford. "Here we are a quarter of a million miles from earth and we are still studying"—the flight plan for lunar approach.

Television converted the world into an international classroom in which many millions were learning about astronomy, technology, and communications; in fact no end of fascinating subjects and lessons were being taught on the blackboard of the universe with the astronauts as teachers. Scientists, geologists, astronomers, space experts, all were in the audience intent on studying the moon's geography and particularly landing sites. Every word and picture were under study. For the layman, excellent animations and explanations by the commentators, assisted by experts in astronomy as well as specialists in study of the moon and its history, added to the value of space communications that invited all to participate in the epochal event.

When about to show the last telecast of the earth, Stafford said that the Apollo crew felt very strong about sharing some of the adventure, the excitement, and the challenge that had been a reward of their eight days in space. "Through this endeavor," he added, "we hope that we have made you and millions of people more a part of history that's being made in our day and age."

Scientists and spacemen profess vital interest in mascons, those parts of the moon where gravity is greater and might be a hazard to spaceships. Lunar photographs and TV are interpreted as making the moon appear more like the earth than was believed. Therefore, astronomers and geologists say that by studying the moon more may be learned about the earth as well as the solar system.

One who remembered listening in on historic events in the pre-television decades of the twenties and thirties could recall no broadcast as exciting as the swoop of the astronauts in the lunar module to 9.4 miles above the Sea of Tranquility—the nearest man had ever been to a celestial body. The radio script for this performance was "written" while being enacted; the keynote was immediacy. While the spacemen conversed among themselves and with Mission Control, the power of the spoken word enlivened by the excitement in the astronauts' voices revealed that radio had lost none of its miraculous ability to convey human emotions and "paint" pictures in the mind even though across 250,000 miles.

Now even in the era of television, radio demonstrated it is an integral part of modern communications, indeed the basic fabric. Radar too distinguished itself in the reconnaisance and reconnoitering; it was the infallible guide. Indeed, this drama booked as "Apollo 10 Journey to the Moon" was far and away from the lone role that

radio played in the Coolidge era when broadcasting was called a
"craze." With remarkable clarity radio afforded listeners everywhere
opportunity to earwitness news and history in the making. It reminded oldtimers of the days when sound alone described the take-
offs and landings of Lindbergh, Byrd, the NC-flying boats, and other
pioneer flights across the Atlantic. Now television as well as radio
from near the moon made those pioneering feats in aviation appear
short-distance.

IMMEDIACY IN EXPLORATION

Science and communications created an epoch of immediacy in
exploration. As soon as an explorer makes a discovery all the world
sees and hears about it. Not always so. Columbus, Magellan, Drake,
De Soto, Peary, Scott, Shackleton, Amundsen, and all the pioneer
trail blazers were lost in silence, mystery, and loneliness for months,
even years. Now, in marked contrast, Cernan, an explorer in space,
excitedly calls out from less than 10 miles from the moon, "We're
right there! We're right over it! I'm telling you we are low, we're
close. This is it!"

And that act in the drama ended only when the 15-ton lunar
module rejoined the command module, superbly navigated by John
Young. With the 31 moon loops, landmarking, 3-dimensional photography, and other tasks completed, the main engine fired to start
Apollo on the 54-hour, 246,154-mile voyage back to earth as
the nineteenth and final colorcast showed sensational pictures of the
receding moon.

A colorful dawn added to the effectiveness of color TV in portray-
ing the splashdown 440 miles east of Pago Pago, American Samoa,
where the carrier U.S.S. "Princeton" was stationed for the recovery
operation. Soon after the almost four minutes of blackout at re-entry,
radio voices from the helicopters reported sighting the capsule. TV
cameras on the ship focused on it as a red dot with a long, long tail; a
small, glowing comet that streaked across the screen. Then came the
sonic boom as the spacecraft broke the sound barrier, and the big
orange parachute spotlighted by the sunrise floated downward toward
the sea. Flashing blinkers on the helicopters gave them the appearance of large fireflies twinkling in the dawn's dim light as they

hovered over the impact spot illuminated by a grenade dropped from one of the copters.

The "Princeton" less than 4 miles away quickly neared the scene, affording the TV cameras magnificent close-up views of the bobbing capsule and the lift of the spacemen to the hovering helicopter. A half-mile flight put them on the deck where the crew welcomed the exuberant astronauts in traditional manner as television recorded the event for national as well as international viewing via satellites. Thus ended a triumphant 192-hour journey of 500,000 miles that established new records in flight, exploration, photography, and communications.

A COUNTDOWN OF ACHIEVEMENT

The valiant Mercury, Gemini, and Apollo men dauntlessly performed in space as testimony of man's vision and courage; they etched their names and American glory in the history of exploration and communications. They were acclaimed for reaching greatness by reaching out beyond themselves.

Rigorously they trained, navigated, and explored in realms designated as seedbeds of research, invention, and engineering. Their endeavors were a spur to productivity, a stimulus to employment, a potential bulwark of national security, a motive for learning, and an inspiration to youth. Their cameras portrayed fantastic art in photography. The cosmos became a fertile region for cultivation of new ideas and thousands of technical innovations, or "spin-offs" of endless import.

Basically significant, the space projects reveal how technology under masterful control and concentration of effort can create an abundance of new materials, devices, systems, and knowledge that spearhead the future. Ingenuity, vision, planning, and mobilization of manpower together with national and scientific resources are keystones of success. Meticulous attention to detail is devoted to every component; a tiny transistor, an electronic switch, or a fuel valve are as important as the giant rocket itself. As a seminar of progress the Apollo Project taught how to organize a complex enterprise, and demonstrated how massed effort of thousands of technical minds and diverse talents can gain unprecedented goals in cooperation with government, industry, and universities.

Communications in every form on earth and in space, reinforced by automation and the computer, successfully solidified an immense world-wide organization that operated like clockwork at peak efficiency in the achievement of what was proclaimed as man's nobler aims and motives. As he extended his domain, radio, television, radar, and telemetry attained new dimensions in performance, range, and reliability surpassing the most sanguine dreams of the Marconi era.

The stage was set; the mammoth Saturn moon rocket was fueled on the launch pad. Computerized inspection showed everything A-OK. The countdown was about to begin for Apollo 11's momentous mission.

VIII Epic Signals from the Moon

> That's one small step for a man, one
> giant leap for mankind.
> —*Neil A. Armstrong*

THE CLOCK-LIKE CADENCE OF THE COUNTDOWN FOR APOLLO 11 WAS
the overture for one of the most thrilling performances in the annals
of electronic communications from the standpoint of history, technology, and histrionics. For veteran television viewers who had
looked in on memorable events since the advent of the art, the
landing on the moon topped them all. TV made the moon into a
theatre, whereas before all the world was deemed to be the stage.

From the awesome instant of blast-off on July 16, 1969, Apollo
11's voyage to the moon for a landing in the desert-like Sea of
Tranquility was a dramatic show packed with excitement and suspense. Man was definitely on his way to stamp multi-billion-dollar
footprints on the moon.

All the world in tune with television saw this Apollo, like its trail-
blazing predecessors, hurtle into space wrapped in smoke, steam, and
terrifying orange flame as it lunged upward from the Florida sands
and out over the Atlantic. Three courageous Americans were on
board for a journey destined to be ever memorable in the history of
exploration, flight, and communications: Neil A. Armstrong, a civilian, and two Air Force officers, Colonel Edwin E. Aldrin, Jr., and
Lieutenant Colonel Michael Collins.

Their spectacular lift-off was relayed via communications satellites
over the Atlantic and Pacific to about 50 countries on six continents.
The vast audience was incalculable, yet those bold enough to guess

estimated a total of 25 million persons, and looking ahead they projected the size of the audience at landing time would break all records and total 500 million, possibly a billion.

After two loops around the globe at an altitude of 118 miles for 2½ hours, the Saturn's third stage rocket refired and sent the Apollo into translunar trajectory at an initial speed of 24,200 miles an hour. "You're on your way now," radioed Mission Control.

Apollo was rushing on a precise course toward its target, some 240,000 miles from earth. Seventeen tracking stations linked by millions of miles of wire and radio to the Spacecraft Center in Houston, Texas, never let the Apollo "out of sight." Communications were loud and clear; in fact, it was difficult to believe that the astronauts were already so far away. Their initial colorcast during the first evening in space proved that the spaceship was swiftly moving away from the receding earth.

The second telecast from 149,000 miles showed the half-earth streaked with clouds, and excellent pictures of the spacemen as they scanned close ups of the instruments and explained the functions.

The third show featured an inspection of the lunar module; the earth was 177,000 miles away and the moon 73,000 miles. As Aldrin looked out the window, he observed, "Everything's getting smaller as time goes on." The miracle of television was evident in every picture; so was its phenomenal educational value.

After rocketing into lunar orbit, they staged the fourth telecast as the astronauts, sailing from 61 to 172 miles above the moon, conducted a remarkably clear tour of the craters and the landing "field" their eyes were fixed upon.

The next big step for the LM, code-named "Eagle," was to separate from the "Columbia" commandship named after the lunar ship "Columbiad" in Jules Verne's 1865 novel *From the Earth to the Moon*. After breakfast on the eleventh revolution of the moon, Aldrin and Armstrong crawled through the tunnel into the LM to prepare for descent. Near the end of the twelfth orbit the vehicles were unlocked, and Armstrong radioed, "Eagle has wings."

Telemetry recorded it was exciting for the astronauts as well as for earthlings. At the time of the descent rocket ignition Armstrong's heartbeat, normally 77 a minute, jumped to 110 and at touchdown 156.

Collins continued to taxi the commandship around the moon at an altitude of about 60 miles, awaiting the return of his fellow astronauts from their epochal adventure. He ran his record for duration in lunar orbit up to 59 hours 27 minutes 55 seconds. The lone circumnavigator had no TV on his solitary trips, but he listened in by radio and like a bleacher fan applauded every move—fantastic!

As the LM descended, radio maintained distinct contact between the astronauts and Houston while TV's graphic simulations and informative commentaries realistically made video-earthlings in their front-row seats feel that they were out in space as eyewitnesses. Revolving figures on the screen displayed the countdown of altitude as the LM dropped about 129 feet a second. Four minutes after the firing, it was down to 40,000 feet; at 7,200 feet the landing site was only 5 miles ahead with the computer in control.

At about 300 feet Armstrong radioed that he had taken over the semi-manual control to guide the Eagle over a boulder-filled crater to a smoother area. And so it is logged that on July 20, 1969, at 4:17:40 P.M., Eastern daylight saving time, Armstrong flashed, "Houston, Tranquility Base here. The Eagle has landed."

About 6½ hours later the hatch opened. On his way down the ladder Armstrong pulled a lanyard that opened the MESA (Modularized Equipment Storage Assembly) thereby exposing the TV camera to scan the remainder of his descent, and his foot first touching the moon. It was 10:56:20 P.M., Eastern daylight saving time, when he remarked, "That's one small step for a man, one giant leap for mankind." With that memorable statement as a prologue, the moon show was under way. Aldrin was seen coming down the ladder, and then the camera was erected on a tripod to televise the area surrounding the LM as two star actors began to observe the strange ghostly sights. With the enthusiasm of tourists photographing Niagara, they were seen snapping photographs as they filmed "the magnificent desolation," as Aldrin described the weird panoroma.

A FLAWLESS PRODUCTION

The production was flawless; a playwright's delight. Showmanship was perfect from the minute that the lunar module dropped slowly like a big bug upon the boulder-strewn, crater-pocked stage. The scenery was expansive and eerie. For the first time in history two men stepped upon the moon to explore the lunar crust. As they peered through the visors of their strange-looking helmets, not a soul was in sight on the moonscape to cast an eye on them and behold their adventurous arrival. Nevertheless, far across the void, hidden in the distance, millions if not billions of eyes were watching as they roamed over the bleak, rocky soil to conduct scientific tasks and collect samples of the terrain.

Television kept its electronic eye on them, and projected their spectral images to earth. They walked into family circles like characters depicted in science fiction, but this scientific visitation was not fiction: it was from real life and live from the moon. It was unbelievable! Fascinated by it all, many a boy and girl, too, decided they would like to go to the moon. Grandpa and Grandma confessed they had never seen anything like it. In youth they had read Jules Verne but never dreamed they would see Americans walking on the moon. Grandpa remembered his father telling how a teacher in the 1860s caught him reading the Verne book and confiscated it as nonsense, nothing for a schoolboy to waste time on.

Now, 100 years later, a "National Day of Participation" was proclaimed "to mark the moment of transcendent drama"; schools were closed so that boys, indeed all Americans, could have the opportunity to see the Verne tale come true on television.

"In past ages," said President Nixon's proclamation, "exploration was a lonely enterprise. But today the miracles of space travel are matched by miracles of space communications; even across the vast lunar distances, television brings the moment of discovery into our homes and makes us all participants."

Like all outstanding plays, whether Broadway hits, operas, or Westerns on the screen, the Apollo show had all the elements playwrights seek and strive to create. Interwoven between every line was spontaneity, mystery, suspense, tension, risk, and menace. Anything might happen at any moment. Menace was ever-present: cosmic hazards, the danger of mechanical or electrical malfunctions among the two million skeletal components and 15 miles of wire in the commandship alone. And, of course, there was chance of human error. Fear lurked "in the wings" that some strange lunar germs might be picked up and if brought back to earth might endanger life and turn the globe into a barren orb to match the moon. An added ominous element was Luna 15, an unmanned spacecraft buzzing around the moon to cause fear of collision, although the Russians maneuvered it out of the way. Tension was relieved when the orbiting craft presumably crash-landed in the Sea of Crises about 500 miles from Tranquility.

While all this was going on, electromagnetic waves traveling at the speed of sunlight—186,000 miles a second—leaped the historic pictures and radio sounds to earth in 1.3 seconds. That fit perfectly with TV's penchant for immediacy.

It almost seemed as if television had handed high-resolution binoculars to everyone on earth. The pictures were that clear as people everywhere, whether they understood English or not, stared in amazement and applauded the incredible drama enacted some 241,500 miles out in space, performed as a show-of-shows.

Uncounted millions of viewers saw television at its best, in one of its finest hours linked with news and history. They heard the older science and art of radio surpass many of its records of years gone by. Radio too had one of its grandest hours. Realistically it enlivened the spectacular TV pictures with the voices of the actors as they conversed, and as they talked with the earthly Control Center. Based on the clarity of radio and ease of the two-way conversations, it almost seemed that the moon men were in Texas too. They had to use radio for man-to-man communication since there is not a breath of air on the moon for sound to set in motion; in fact, they couldn't hear a spoken word or even a shout in the open.

As they trudged, shuffled, and even "kangaroo-hopped" in their bulky, back-packed space suits, the TV eye roved with them. Their dialogue was so natural that one might think they had rehearsed their lines before leaving earth. They were seen proudly unfurling Old Glory on the moon and then installing scientific instruments to be left there for remote experimentation.

Modern magic in communications made it look simple and easy. With the nearest telephone booth a quarter of a million miles away, it seemed strange to hear a voice from the Control Center inform the spacemen that the President of the United States was "on the wire" to talk with them and to express the nation's congratulations. Split-screen TV technique showed Richard Nixon at a telephone in the White House and the astronauts standing near the 23-foot-high LM alongside the American flag, while they conversed as if just across the Potomac.

The Control Center kept them informed on the clock and warned that the time was approaching for them to climb back into the Eagle. Their extravehicular activity having been completed, TV pictured them phantom-like going up the ladder; they had been afoot on the moon 2 hours, 21 minutes. The "chalky gray" terrain "pretty much without color," which had been their stage, was littered with props and gear discarded to conserve weight, including their portable life-support systems, sample-gathering tools, solar energy panels, boots, and the LM descent stage, as well as scientific instruments. Amidst it all was the TV camera, lifeless on its tripod, and the Stars and Stripes.

TIME TO GO HOME

After spending 21 hours and 36 minutes on the moon, it was time to fire the ascent engine for lunar lift-off.

Mission Control radioed, "You're cleared for take-off."

"Roger, understand. We're Number One on the runway," replied Armstrong, who shortly afterward radioed, "Right down U.S. One."

The rendezvous radar looked spaceward to track the command-ship, and provide the astronauts and their guidance system with the important data necessary to accomplish a successful dock. Within 3½ hours the Eagle nudged up to the Columbia, reuniting the crafts as Apollo 11. Since there was no TV to cover the blast-off and docking, it remained for radio between the astronauts and Houston as well as colorful animations on home screens to reveal the miracle of radar that reunited Armstrong and Aldrin with Collins in the command-ship.

It was not long before they fired the main rocket engine to take them out of lunar orbit and free them from the grip of the moon's gravitation. They were on the way home, having performed an unprecedented mission without a hitch. The famed Eagle was jettisoned, cast adrift in space but not likely to be abandoned in the memory of those who witnessed its triumph.

Coasting along at increased speed under influence of the earth's gravitation, the fifth colorcast portrayed the receding moon, which the astronauts pointed out as the place they had been, and as the camera focused on the half-earth 175,000 miles away, they said, "That's where we are going." Collins conducted several demonstrations of weightlessness, and Armstrong called attention to the two boxes of lunar rocks in the valued cargo.

During the sixth and final colorcast, as Apollo raced down "U.S. 1" at accelerating speed, the earth-bound astronauts thanked the multitude of American workers, scientists, engineers, and communicators who had participated in the many phases of the complex project. Collins noted that the trip "may have looked simple and easy," but he added, "I want to assure you that this has not been the case—All you see is us but below the surface are thousands and thousands of others." Aldrin agreed it was "far more than a three-man voyage—a symbol of the insatiable curiosity of all mankind to explore the unknown."

With those expressions of appreciation the astronauts signed off on TV and began preparation for return to earth the next day.

SPLASHDOWN!

The recovery scene on July 24—the eighth day of the flight—was similar to the other Apollo splashdowns, and communications were conducted with the same precision and efficiency. It was dawn in the Pacific about 950 miles southwest of Hawaii, where the carrier U.S.S. "Hornet" was stationed with a flock of rescue helicopters. A thick cloud cover and the dim light of predawn prevented TV from spotting the parachutes, but radio carried the cheers of the crew when visual contact was established and someone called out, "They're back from the moon." Flight time was logged at 195 hours, 18 minutes.

The seared capsule dropped within a mile of the planned target; the "Hornet," 9 miles away, shifted course and steamed at 22 knots to the point of impact. When the ship edged up within 4 miles, the TV cameras sighted the spacecraft tossing on the waves and the copters overhead. Hoisting the astronauts from the raft to the recovery copter was the signal for a Navy band aboard the carrier to strike up "Columbia, the Gem of the Ocean."

Within a few minutes copter 66 alighted on the deck. This time, however, there was no red carpet, no gold-braid reception line-up on the deck for a welcoming celebration. Quickly the copter was lowered to the hangar bay area where a Mobile Quarantine Facility was set up. Only for a minute were the heroes seen as they stepped from the copter into quarantine, garbed in isolation garments delivered to them by frogmen. For 21 days they would be in quarantine as all precautions were taken to guard against any contamination, germs, or organisms that might have been picked up on the moon.

Nevertheless, it was a hero's welcome that spread around the globe. Added interest was sparked by the presence of President Nixon on board the "Hornet." After watching the recovery operation he went to the window in the door of the quarantine trailer where without direct contact he enthusiastically chatted with the trio of happy astronauts, shown by TV to be in excellent condition. The ship's chaplain delivered a prayer of thanksgiving, and following the benediction the band played "The Star Spangled Banner." The natural wonder of television was evident at every turn as the historic pictures were beamed to a communications satellite above the Pacific for relay around the world.

As the President waved good-by, the curtain on the window of quarantine was drawn. TV followed him to the flight deck, where he boarded a helicopter and took off into the east to continue a worldwide tour. Thus the TV curtain was lowered on the "Hornet," ready to sail for Pearl Harbor with the historic Columbia, along with the astronauts confined in the Quarantine Trailer.

Television covered the Honolulu welcome, while no time was lost in hauling the spacecraft to nearby Hickham Field for an eight-hour flight to the Space Center at Houston; the treasured boxes of lunar specimens went that way too. The isolation trailer was put aboard an Air Force C-141 cargo plane and flown to Texas, where the astronauts would spend the rest of quarantine. TV was on hand in the middle of the night to picture the welcome home and through-the-window reunion with families and friends.

Standing by were planetologists, geologists, biochemists, and scientists in diverse categories anxious to begin exhaustive tests, hopeful for clues in the moon rocks that might help solve some of the age-old mysteries of the moon, and possibly the earth.

No letup was in sight for the TV crews. A triumphant greeting by a joint session of Congress was televised as were national celebrations and gala parades. The astronauts then left on a 22-nation tour.

INNOVATIONS IN COMMUNICATIONS

Communication systems aboard Apollo 11 featured all the advances used so effectively in previous Apollo flights and in addition "innovative accomplishments" in Space Age electronics related to radio, radar, telemetry, and television.

Two TV cameras built by Westinghouse were on board, one for color and the other black-and-white. The 13-pound color camera with a variable-focus "zoom" lens televised the astronaut activity in the commandship enroute to and from the moon. The smaller black-and-white 7.25-pound camera featured some 250 microminiature integrated components and circuitry. Another innovation—a "mini" monitor with a screen about the size of a credit card enabled the astronauts in the Columbia to see in black and white the pictures being transmitted to earth.

Engineers called attention to the fact that the Apollo color camera designed for low light level operation, used only one imaging tube. Technically the method is known as "field sequential." A disk about 3 inches in diameter, divided into six sections containing red, blue, and green filters, spins at 600 revolutions a minute to pass the sequence of colors in front of the imaging tube. In effect, the camera takes many color pictures at a rapid rate. In this case, as each color filter passed in front of the imaging tube, it collected all the information related to red, blue, and green and continued to repeat the sequence. Each picture was transmitted to earth where special conversion apparatus combined them like a matrix, one on top of the other, to produce a multicolored image on the screen.

In contrast with the sequential system, standard commercial TV cameras use three imaging tubes, each with a separate color photographic function. One concentrates on red, the second on blue, and the third on green. When electronically combined they produce one picture of various shades, hues, and colors on home screens. That is called the simultaneous method, which has the great advantage of being compatible with black and white; it can show the picture in either color or monochrome.

In spatial television much electronic magic intrigues the eye in a split second, and in a very roundabout way. For instance, TV signals beamed from the moon were picked up by a 210-foot radio-telescope dish antenna at the National Astronomy Observatory, at Parkes, Australia. From there, the signals converted into standard television pictures were microwaved to Sidney and then via an Intelsat communications satellite above the Pacific to the Mission Control Center in Texas for distribution to the national networks and relay overseas.

As if to challenge the ingenuity of the electronic "magicians" as much as possible, the Apollo black-and-white lunar camera transmitted pictures composed of 320 lines per frame at 10 frames a second instead of the conventional 525 lines per frame at 30 frames a second. That was done to conserve power and communication bandwidth. The variation necessitated scan conversion on the ground, otherwise the pictures would flicker badly. That was overcome by scan converters featuring an instantaneous recording play-back process similar to that used for "instant replay" in sports telecasts. The converter installations were in California, Spain, and Australia where the transformation or conversion of the pictures to conventional TV standards used by the networks took place in the twinkle of an eye and unbeknown to those looking in.

SCIENTIFIC PACKAGES

When the Eagle lifted off the moon, the TV circuits were severed by separation of the ascent and descent stages; and the lunar camera with power near the diminishing point was left behind sightless and mute.

Also relegated to the moon were two measurement-making instruments—a Passive Seismic Experiment Package (PSEP) to measure moonquakes, meteoroid impacts, or any other tremors. Placed nearby on the rocks was a Laser Ranging Retro-Reflector (LRRR) for precise measurements of the distance between the moon and earth; first tests showed the moon about 131 feet farther away than previously figured. LRRR is expected to operate for ten years as a target for laser beams aimed from the earth. PSEP was designed to function for about a year.

Those two experiments constituted the Early Apollo Scientific Experiments Package (EASEP) of NASA, to be succeeded by ALSEP, the Apollo Lunar Surfaces Experiments Package, to be placed in future landings as a sophisticated geophysical lunar station. Eventually ten such stations are envisaged in operation, transmitting data relative to the moon's physical properties and electromagnetic characteristics. When received on earth the incoming information will be entrusted to computers in the hope that they may project conclusions on what the moon is actually like. For example, prove or disprove whether the lunar core is liquid or hot; is it homogeneous and possibly a mass formed in outer space and ensnared in the earth's gravity, or is it composed of concentrations of mass caused by heat and volcanic activity?

RADAR'S UNCANNY VISION

Radar gave the astronauts long-range and precise electronic vision. It told the LM pilots where they were in relation to the command-ship, and provided important data for the spacecraft's guidance system, so that the proper engine firings could be achieved to bring the Eagle to a rendezvous with Columbia when returning after the lunar landing.

As the LM approached close to the lunar surface, the landing radar went into action, bouncing four radio beams off the moon to inform Armstrong and Aldrin their distance above the moon and how fast they were approaching it—similar to data provided by an aircraft pilot's altimeter and speed indicator. The landing radar information like that of the rendezvous radar was flashed before the pilots and fed into the LM's guidance system so the descent engine could slow the spacecraft to a soft landing, like a helicopter hovering before landing.

RADIO HIGHLIGHTS

The entire communications system designed to span a quarter-million miles of translunar space, and to transmit and receive voice signals, telemetry data, astronaut biomedical information, tracking signals, and television weighed just under 100 pounds.

When Armstrong and Aldrin stepped out of the LM to explore the moon they carried in their portable back-pack life-support system a cigar-box-size RCA Communications unit with which they talked to one another over Space Age versions of walkie-talkies; the microphones were tucked in their helmets. The LM radio installation acted as a radio relay station with mission controllers on earth. Although each back-pack radio contained five transmitters and receivers plus telemetry instrumentation to radio data on the astronauts' physical

FROM MARCONI TO MAN ON THE MOON

Guglielmo Marconi and his aides, Kemp (left) and Paget, at Newfoundland in 1901 received the first transatlantic wireless signal. The reading world was skeptical; not so 68 years later when Apollo 11 carried Neil Armstrong, Michael Collins, and Edwin Aldrin to the moon. TV's dramatic portrayal left millions of watchers no room for skepticism, only belief and fascination.

Astronauts' photo: *NASA*

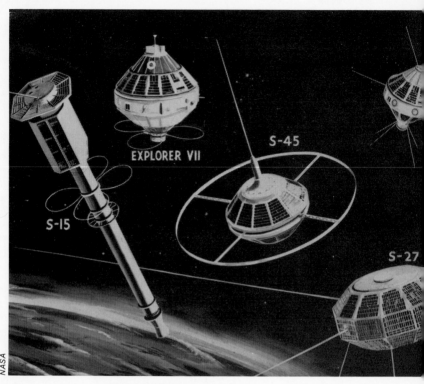

A FAMILY OF SATELLITES *(Typical Goddard Missions)*

S-15   Gamma Ray Telescope
S-45   Ionosphere Beacon
S-48   Topside Sounder (U.S. Scout XI)
S-27   Topside Sounder (Canada)

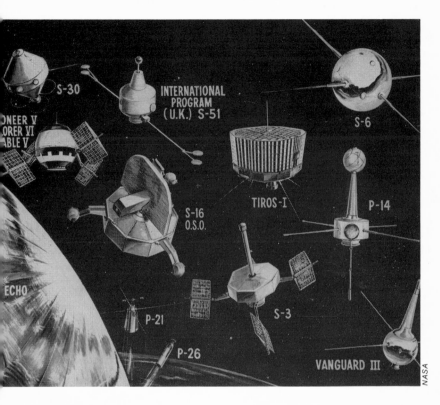

NASA

S-30	Ionosphere Direct Measurements	P-14	Magnetometer Probe
S-6	For Study of the Atmosphere	P-21	Electron Density Profile Probe
S-3	Energetic Particle Satellite	P-26	NERV (Nuclear Emulsion
S-16	Orbiting Solar Observatory		Recovery Vehicle)

Grumman Aircraft

MAKING HISTORY IN COMMUNICATIONS

The flight of Apollo 11 to the moon in July, 1969, established epochal records in communications—first radio and television from the lunar surface. At about 70 miles above the moon the spidery-legged module (LM) separated from the command ship and descended to land Armstrong and Aldrin at Tranquility Base. Collins continued to pilot the "Columbia" around the moon awaiting the ascent

Jet Propulsion Laboratory

Mariner 6 and 7 spacecraft flew by Mars in July-August 1969 and across 58 million miles radioed photographs of that planet in greater detail than man had ever seen.

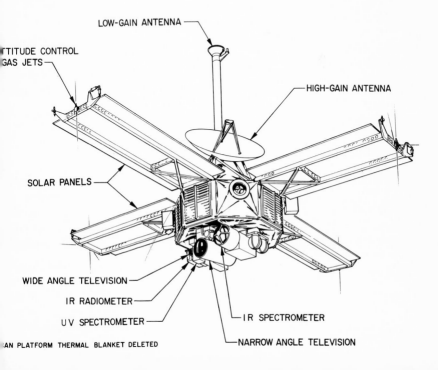

LOW-GAIN ANTENNA

ATTITUDE CONTROL
GAS JETS

HIGH-GAIN ANTENNA

SOLAR PANELS

WIDE ANGLE TELEVISION

IR RADIOMETER

UV SPECTROMETER

IR SPECTROMETER

NARROW ANGLE TELEVISION

SCAN PLATFORM THERMAL BLANKET DELETED

HOW TIMES HAVE CHANGED

Marconi at his transatlantic "control center," Glace Bay, Nova Scotia, in 1902.

The activity of scientists, technicians, and electronic equipment in the NASA blockhouse at Cape Kennedy during operations of the Mercury Redstone rocket that placed astronaut Alan B. Shepard, Jr., into a 5,100-mile-per-hour suborbital flight on May 5, 1961.

Members of the government-industry team at the Launch Control Center, Cape Kennedy, stand at their console monitors for a direct window view of the Apollo 11 blast-off on July 16, 1969.

ECHO BALLOON SATELLITE

This "passive" satellite, a huge aluminized plastic balloon, was used successfully by scientists at Bell Laboratories to bounce radio signals, telephone conversations, and radiophotos between New Jersey and California as it streaked across the sky 16,000 miles an hour in an orbit 1,000 miles high. The "big ear" horn-shaped antenna, 50 feet long, rotates to track fast-moving satellites and to scoop up communications relayed from the sky.

Bell Telephone Laboratories

Hughes Aircraft

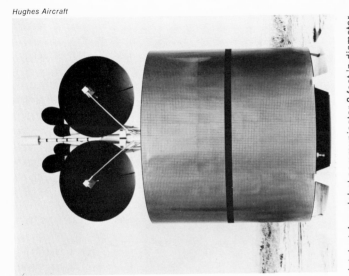

Intelsat 4, a global communicator 8 feet in diameter and 17½ feet high, is designed with a unique antenna system to carry 5,000 two-way telephone calls or 12 TV programs.

Syncom "parked" in orbit 22,300 miles up was launched on July 26, 1963, as the first synchronous communications satellite within range of "seeing" 40 per cent of the world.

NASA

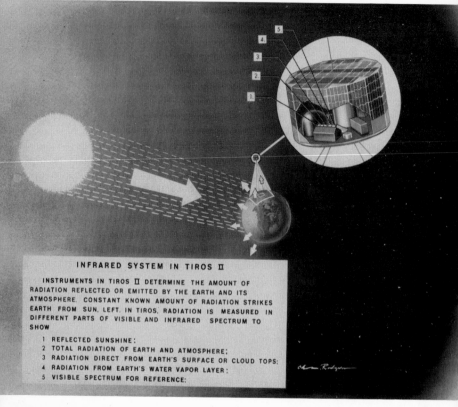

INFRARED SYSTEM IN TIROS II

INSTRUMENTS IN TIROS II DETERMINE THE AMOUNT OF
RADIATION REFLECTED OR EMITTED BY THE EARTH AND ITS
ATMOSPHERE. CONSTANT KNOWN AMOUNT OF RADIATION STRIKES
EARTH FROM SUN, LEFT. IN TIROS, RADIATION IS MEASURED IN
DIFFERENT PARTS OF VISIBLE AND INFRARED SPECTRUM TO
SHOW

1 REFLECTED SUNSHINE;
2 TOTAL RADIATION OF EARTH AND ATMOSPHERE;
3 RADIATION DIRECT FROM EARTH'S SURFACE OR CLOUD TOPS;
4 RADIATION FROM EARTH'S WATER VAPOR LAYER;
5 VISIBLE SPECTRUM FOR REFERENCE;

WEATHER SATELLITE

Tiros and Essa satellites equipped with TV cameras have sent back thousands
of pictures of cloud formations and hurricanes as well as scenes of the earth
viewed from orbit.

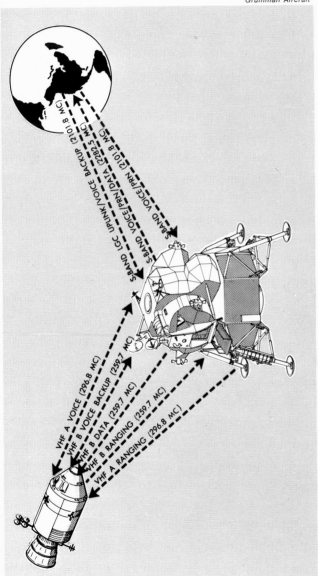

Grumman Aircraft

APOLLO 11 IN-FLIGHT COMMUNICATIONS

Communications systems used by the command ship "Columbia" and the lunar module (LM) "Eagle" for contact with each other and the Control Center on earth, as well as for television, radar, and telemetry while in flight and from the lunar surface.

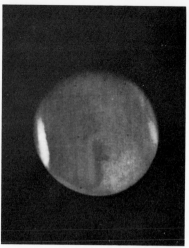

Jet Propulsion Laboratory; Mars photos, NASA

FOCUSED ON MARS 60 MILLION MILES AWAY

A 210-foot antenna at the Goldstone Tracking Station near Barstow, California, in conjunction with a high-rate telemetry system, picked up pictures of the ruddy planet transmitted by Mariner spacecraft at the rate of 16,200 bits a second.

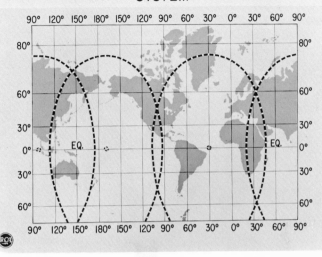

WORLD COVERAGE WITH 3 SATELLITE SYNCHRONOUS SYSTEM

Orbiting 22,300 miles above the equator, a communications satellite completes one trip around the globe in 24 hours while the earth completes one full rotation. To a ground observer the satellite seems permanently fixed above one point of the earth's surface. When the satellite and earth are in step, the system is called synchronous.

A synchronous satellite system.

22,300 MILES

Bell Telephone Laboratories

EVOLUTION OF A SATELLITE

Telstar, orbited in 1962, pioneered a new era in long-distance communication and established memorable milestones in transatlantic radio, telephony, and television.

Bell Telephone Laboratories

Discharge tube of a metal-vapor laser, rated as one of the most efficient and least expensive continuous-wave lasers. Cadmium vapor is used for ultraviolet and blue lasers, tin vapor for red, and zinc for infrared.

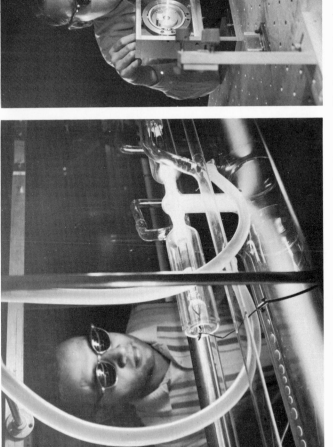

The laser's coherent light beams demonstrate great promise for use in communications. The high-powered gas laser, pictured here, mixes helium, carbon dioxide, and nitrogen. It has produced continuous outputs of more than 106 watts.

Bell Telephone Laboratories

Bell Telephone Laboratories

THE "BIG EAR" ANTENNA

Giant horn-shaped antenna built by the Bell System near Andover, Maine, for overseas satellite communication experiments. It is protected from the weather by the largest inflated radome yet made—210 feet across and 161 feet high, made of synthetic rubber and fabric "transparent" to radio waves.

condition as well as the status of their space suit systems, it weighed only 6.5 pounds and measured 1¼ by 6 by 14 inches. In case of a critical development such as malfunction of the oxygen supply or cooling equipment in the space suit support system, radio would sound a special tone to warn the astronaut.

The compact Extra Vehicular Communications System (EVCS) was designed so that when both spacemen were outside the lunar module Armstrong served as a relay to the LM for Aldrin. He transmitted FM (frequency modulation) signals bearing both voice and telemetry to Armstrong's set. In turn, Armstrong's set combined his voice and telemetry with that from Aldrin and beamed everything over AM (amplitude modulation) to the LM for automatic relay to earth. Collins kept in touch through the LM-earth-CSM relay link.

Briefly that outlines the complex electronic communications instruments and systems that contributed so magnificently to one of the greatest technological achievements of all time—opening of the Lunar Age!

LUNAR ENCORE

An all-Navy crew of Commanders, Charles Conrad, Jr., Alan L. Bean and Richard F. Gordon, Jr., shoved off on November 14, 1969 on the second journey to the moon. In flight and in communications their spaceship was essentially the same as Apollo 11 from blast-off to splashdown except for several unexpected incidents.

To begin with, Apollo 12 lunged upward in a harrowing lift-off from a rain-soaked launch pad and rocketed into weather described as the worst ever encountered by an American spacecraft. Within 35 seconds after take-off, a mysterious electrical jolt believed to have been caused by lightning triggered alarms, lights and circuit-breakers in the cockpit.

Apprehensive moments ticked off in communications from the instant Conrad radioed, "We had everything in the world drop out," until he pronounced "Everything is tickety-boo—we're on our way." Inspection by the spacemen and analysis of the blitz by the ground controllers led them to decree, "You got a go orbit." That proved to be true several hours later when a colorcast showed the earth 5,000 miles away.

Early in the morning of the sixth day out from earth, television pictured the lunar module Intrepid pulling away from the command-ship Yankee Clipper for the swoop-down to the Ocean of Storms. Radio reported Intrepid landing on schedule and on target with a timetable of approximately 32 hours on the moon. The color TV camera mounted inside the equipment bay showed Conrad and Bean backing down the ladder and imprinting footsteps in the soft, dusty soil. After about 44 minutes when the camera was being shifted for placement on a tripod, it went blind depriving earthlings from watching the explorers as they took two strolls totaling 8 hours.

Radio continued to cover the story as the astronauts exuberantly chatted and conducted colloquy with the ground controllers at Houston. On TV animated simulations and informative commentaries were substituted for the live telecasts that would have pictured the men walking on the moon, planting the American flag, collecting rocks and deploying instruments from (ALSEP) Apollo Lunar Surface Experiments Package. The scientific emplacement included a seismometer, a lunar ionosphere detector, a solar wind spectrometer and a magnetometer.

The heart of the array was a central radio station linked to all the instruments by a ribbon-like electric tape laid on the lunar crust. The unique set-up, virtually an automatic scientific observatory designed to function for at least a year, was powered by a SNAP-27—a 63.5 watt radioisotope thermoelectric generator developed by General Electric for the Atomic Energy Commission; in the process heat generated by radioactive decay of plutonium 238 converted into electricity.

On the seventh day of the mission, Conrad and Bean climbed out of their lunar outpost for a second walk to gather more rocks and soil, snap pictures and hike over to Surveyor 3, a man-made landmark on the edge of a crater since it landed in April, 1967. The pinpoint touchdown of Intrepid put it about 600 feet from Surveyor, parts of which were retrieved including the TV camera for examination before becoming a museum piece. Aerospace scientists were anxious to learn how the robot's hardware survived in the lunar environment. While the astronauts gathered that evidence and completed other assignments, Gordon piloted the Yankee Clipper in orbit about 70 miles high and took photographs of the moonscape.

When the commandship swung into its final solo revolution Intrepid blasted upward to rejoin it. Earthlings saw the moonlander floating back into the world of television as Gordon's color camera scanned it coming "right down the pike." Then came the docking maneuver televised as an incredible performance in navigation and communications.

As soon as Conrad and Bean unloaded their cargo including about 90 pounds of rocks and soil, they crawled through the tunnel for reunion with Gordon. The ascent stage Intrepid was jettisoned to crash on the moon. The Yankee Clipper turned homeward after spending a day in lunar orbit photographing sites for future landings. When 108,000 miles from earth the spacemen held the first TV space-to-earth news conference, answering questions radioed from reporters at the Houston Control Center.

The Clipper was speeding toward the target area about 400 miles southeast of Pago Pago, and 2,651 miles southwest of Honolulu where the U.S.S. "Hornet" was standing by. The bright morning sun provided ideal lighting for spectacular television views of the splashdown and recovery on the white-capped Pacific; the "Hornet" was about 4 miles away and moved in to less than 900 yards.

Shortly after the copter landed the spacemen on the carrier's deck they entered the quarantine trailer. President Nixon telephoned congratulations and told them, "We all went with you on television—I can assure you that millions in the United States and around the world were watching." Then, exercising his prerogative as Commander of the armed forces, he promoted each to the rank of captain.

Thus ended the epochal encore to the moon. Historically, 1969 with two unforgettable lunar voyages marked on the calendar became the "1492" of the Space Age.[1]

[1] Future missions to the moon were scheduled by NASA as follows:

Apollo 13: Captain James A. Lovell, Jr., Lieutenant Commander Thomas K. Mattingly, 2nd., Fred W. Haise, Jr., for the spring of 1970.

Apollo 14: Captain Allan B. Shepard, Jr., Major Stuart A. Rossa, Commander Edgar D. Mitchell, for autumn of 1970.

Five additional Apollo missions were projected up to 1975.

IX Pioneering for Progress

> Nature is an infinite sphere, whereof
> the centre is everywhere, the circumfer-
> ence nowhere.
>
> —*Blaise Pascal*

THE SPACE AGE OPENED WITH A BANG IN 1957–1958. THE FIRST satellite Sputnik 1 soared into orbit from the Soviet Union on October 4, 1957. That date is recorded in the annals of space as the opening of a new era in technology, exploration, and discovery; a new epoch in flight and voyaging. Sputnik, a 184-pound sphere, orbited the earth 15 times a day at 18,000 miles an hour, 560 miles up. Its radio peeped continuously so that it could be tracked until the return to earth on January 4, 1958.

In the meantime Sputnik 2, carrying a payload of 1,120 pounds, including a dog Laika, was orbited on November 3, 1957. By the time of re-entry on April 14, 1958, Laika had died in space. Then came Sputnik 3, launched on May 15, 1958, with a record-breaking weight of 2,925 pounds.

Explorer 1, the first United States satellite (30.8 pounds) designed and built by the Army Signal Research and Development Laboratory at Fort Monmouth, New Jersey, was launched on January 31, 1958. It carried 18 pounds of instruments credited with supplying data that led to the discovery of the Van Allen radiation belt.

The Explorer was followed by Vanguard 1, a 3-pound satellite of the U.S. Navy, orbited on March 17, 1958. Measuring 6.4 inches, about the size of a 2½-pound grapefruit, it was the first satellite equipped with solar cells to convert sunlight into electricity. After six years, having traveled 794,304,000 miles in 23,640 orbits ranging from 406 to 2,463 miles above the earth, Vanguard's radio began to weaken, but it had afforded scientists opportunity to make accurate radio fixes that indicated the earth to be slightly pear-shaped. Finally, in February 1965, near the end of seven years, the radio went silent. The craft, however, is expected to continue going around the globe for another 200 years before dipping into the earth's atmosphere to burn up from friction.

Vanguard 2, a 23.3-pound satellite, was launched on February 17, 1959; Vanguard 3, weighing 100 pounds, on September 16, 1959. The Vanguard program—three satellites in orbit out of 11 launchings—was reported to have cost about $100 million.

PIONEERING RECORDS

At the outset the United States established three historic records that showed space communication was practical:

1. Score, a satellite outfitted with radio and a tape recorder that stores voice messages on tape for delayed repeat on command from the ground, was launched by the U.S. Signal Corps by an Air Force Atlas rocket on December 18, 1958. It beamed President Eisenhower's prerecorded Christmas message to the world, demonstrating the first voice communication from space, for twelve days until its batteries went dead.

2. Echo 1, a 100-foot aluminized plastic balloon launched by NASA on August 12, 1960, bounced the first two-way satellite-relay telephone conversations cross-country by reflection of the voice-carrying radio waves from its mirror-like surface.

3. Courier 1B, a delayed-repeater launched by the U.S. Army in October 1960, equipped with solar cells, demonstrated the equivalent of 16 two-way, high-speed teletype channels before it went silent after two weeks.

Who among the radio pioneers ever dreamed that an American satellite such as Pioneer 5, launched in May 1959, would send back radio signals from more than 20 million miles, shattering existing records of long-distance communication.

Since then many spaceships and satellites have been orbited. And of course each had to have a name. Continually, new words are coined and significant names identify the satellites: Advent, Courier, Discoverer, Essa, Eros, Explorer, Mariner, Mercury, Midas, Nimbus, Orbiter, Prospector, Ranger, Relay, Samos, Secor, Slomar, Stomar, Surveyor, Telstar, Syncom, Tiros, Transit, Venus, Voyager, and others. Not to forget Early Bird, Lani Bird, Canary Bird and the Intelsats as "switchboards" in the sky.

The exploring satellites are equipped with all sorts of scientific data-gathering instruments—infrared, nuclear, and electronic. Abetted by the transistor and other semiconductor devices, electronics was miniaturized to meet the new environment in which every ounce is important. Bulky TV cameras used on earth had no place in space; they had to be miniaturized to the size of a water tumbler—and they were.

A tiny, 2-pound TV camera about the size of a home-movie camera, was introduced for use in spacecraft to observe astronauts, or the camera hand held by astronauts could scan space panorama, or be used outside the craft following a moon or planetary landing. Another miniature device, a TV tape recorder-reproducer, was designed to record TV signals during intervals when a spacecraft is outside the range of ground stations. Then, when contact is established, the recorder "dumps" the information eight times faster than recorded.

Scientific devices had to be packed into small spheres no larger than a grapefruit or a basketball. But that was only the beginning. As satellites become larger and automatic instruments can no longer make new measurements, the urge will become greater to send astronauts and scientists along to operate the instruments as the spaceships sail across the heavens unhampered in flight by the earth's gravitational field.[1]

Scientists explain that gravity acts as a brake on a vehicle leaving a planet and as an accelerating force on one approaching a planet. Traveling away from a launching pad on the earth the space-bound vehicle must be accelerated to a given minimum speed to overcome permanently the effects of the earth's gravity. That speed, known as the velocity of escape, is slightly more than 7 miles per second, or

[1] Tiros satellite, to survive launching, was designed to withstand 30 times the force of gravity.

25,000 miles per hour at the earth's surface. Escaping from the earth in a rocket is described as a long uphill climb. All scientific instruments in a missile or satellite obviously must be designed to withstand the rocket's terrific thrust and rigorous conditions of outer space.

"With a space mission, every piece in the system encounters more stress and shock in the few minutes of launching than most ground systems face during years of operation," explains an astroelectronics engineer. "We know too, that the space environment will expose our handiwork to greater extremes of temperature, vacuum, and radiation than anything normally found on earth. Then, when the electronic package is mounted on the rocket and sealed up, we have our very last look at it. If any one of the 20,000 or more parts gives out—even a conventional transistor—no service man will be around to slip in a replacement."

RULES BUT NO GROUND RULES

Engineers point out that in space there are plenty of rules, but no ground rules. And by no means has space proved to be paradise for men or machines. Odd things happen up there.

Ordinary lubricants evaporate in the high vacuum hundreds of miles above the atmosphere. Objects become weightless so that a slight impetus of the wrong kind will cause them to tumble or spin almost indefinitely unless special means are devised to control their motion. Absence of any atmosphere to level off extremes of temperature results in great heat on surfaces facing the sun, and extreme cold on those in the shade. Every bit of electronic equipment, therefore, is buffeted in the laboratory on shock and vibration machines, and baked and frozen in vacuum chambers; every component is checked again and again to make sure that the instruments can withstand assault in the brutal environment that prevails in the infinitely extended regions of space.

A BIG PROBLEM

All of the scientific advances that put space communications on the commercial threshold presented problems of considerable magnitude: how to establish a commercial satellite communications system in the best interest of the public and the nation. Science, research, engineering, finance, and industry, all are involved in the development which ferments with far-reaching political and economic implications, both domestic and foreign.

There is no 12-mile limit along the shorelines of the cosmic sea as in international waters. As *The New York Times* observed, space above the atmosphere is a lawless region governed by no agreed body of rules; and even so elementary a question as where the atmosphere ends and space begins has never been settled. And at present the simple truth is that any nation is free to do what it pleases—within the limits of its technological capabilities—in the regions in which earth satellites orbit. It is vital, therefore, that statesmen, diplomats, and international lawyers formulate and establish space law, possibly through the United Nations.

Frequency allocation for space communications in itself is a complicated problem world-wide in scope. Solution calls for long-range vision to guard against impairment of progress in a spectrum that is fertile and in a system that must be kept flexible for new developments and expansion. Indeed, the history of science leaves no doubt that the future of communications across the kingdom of space will feature swift changes and unprecedented innovations.

IMPORTANT QUESTIONS

What type of satellite is the most practicable? What frequencies or channels are most effective?[2] How many satellites, from 3 to 50, will do a satisfactory job on an international scale? These were only a few of the questions. Scientists said that they too had plenty of technical puzzles that could only be solved in the laboratories and by actual tests in space.

As communication experts continued to study the various proposals for a satellite system they found that cost estimates varied over a wide range from $150 million to $500 million, depending upon the

[2] The International Telecommunications Union conference in the autumn of 1963 at Geneva proposed allocation of radio frequency bands for use by communication satellites. A total band width of 2,800 megacycles divided into four bands in the 500-megacycle region, two in 300-megacycle region, and one in the 200-megacycle region, was calculated to be adequate for growth of satellite communication traffic through 1980.

system and the number of satellites and ground stations necessary. Engineers believed it entirely possible that the system finally adopted might feature a combination of low- and high-altitude satellites and thereby gain the advantages of both.

Two Types of Satellites Satellites are classified as *passive* and *active*. The former merely reflects radio signals from its surface like a mirror; no instruments are aboard. The active satellite is more complex; it carries electronic equipment to receive, amplify, and retransmit signals. Such a satellite, of course, presents problems in complexity and life-span. Engineers know that these outposts must be foolproof in the environment of space (for you cannot get up there to fix them if something goes wrong). They are hopeful for success, however, since more than 15 hundred electron tubes have functioned without failure in the submarine cable system under the oceans for from two to six years. And the satellites have the added advantages provided by transistors and other semiconductors.

It is also interesting to note that the number of satellites essential for continuous communication depends to a great extent on the altitude of the orbit; the higher the altitude the longer the simultaneous visibility and the fewer satellites required. It is no easy job, however, to place a satellite in orbit, and exactly to position and orient it.

A Three-Way Plan Experts of NASA outlined three ways a satellite relay system might be established:

1. A satellite orbiting in a circular path at an altitude of 22,300 miles would travel just fast enough to remain in step with the earth's rotational speed and thereby remain in the same place relative to the earth's surface. It is estimated that three or four of these satellites equally spaced along the equator would suffice to provide communication service to most of the world.

2. Lower orbiting satellites would not stay in the same relative locations over the earth, but the engineers calculate that 36 small satellites up 3,000 miles would assure that one would always be in position between the ground stations.

3. A flock of passive balloon-shaped satellites might also accomplish the trick by reflecting voices off their aluminum-coated plastic skins.

An All-Purpose System In reply to the FCC's request for suggestions, a majority favored adoption of a single, all-purpose satellite system that would be available to all nations for world-wide telephone, radio, television, telegraph, and data services, through two or three relay stations "stationary" in space 22,300 miles above the equator. At that altitude, in an orbit directly over and parallel to the equator, the satellite would take exactly 24 hours to complete a trip around the globe. During the same period the earth completes one full rotation on its axis beneath the satellite; thus to a ground observer, the satellite would remain permanently "fixed" above one point on the earth's surface. For that reason, this system is called synchronous; the satellite's orbital motion is synchronized with the rotation of the globe.

Engineers estimated that such a system utilizing only two satellites, one over the South Atlantic and the other in a similar position over the Central Pacific, could link the major international communication areas of both hemispheres. By using three satellites, the third over the Indian Ocean, it is believed that every inhabited part of the world could be covered with substantial overlaps. Synchronous satellites are depicted as relatively simple, reliable mechanisms that contain equipment to receive, amplify, and retransmit the equivalent of 1,000 or more two-way voice circuits for telephone and other services.

When scientists at the Bell Laboratories heard that NASA planned to launch a plastic balloon-type satellite with an aluminum skin, they suggested that it be used for communication experiments as well as for the original purpose of measuring the density of space, influence of the pressure of sunlight, and other effects. As the giant balloon, Echo 1, passed over the United States, a "big ear" antenna at Holmdel, New Jersey, picked up a radio signal carrying the recorded voice of President Eisenhower, transmitted from the Jet Propulsion Laboratory at Goldstone, California, and bounced off the balloon. Reception was described as being clear as a local broadcast. Continuing the experiments, two-way telephone calls between California and New Jersey were echoed off the balloon from an altitude of about 1,000 miles as it traveled 16,000 miles an hour, encircling the earth every 121.6 minutes. As a further test, a photograph of President Eisenhower sent 1,000 miles into space by an Associated Press wirephoto transmitter used in connection with a big dish antenna at

Collins Radio Company, Cedar Rapids, Iowa, was bounced off Echo and intercepted in Texas.

Signals from the balloon were measured by the engineers as "an energy of about one-billionth of one-millionth of a watt." To dramatize it, they say that if a man were 10,000 miles out in space holding a one-watt flashlight lamp, the amount of light reaching the earth from that lamp would be about equal to the radio energy reflected from the balloon.

Success was attributed in large measure to a supersensitive low-noise, feedback FM receiver developed by Joseph G. Chaffee of the Bell Laboratories in 1933. Applied to earthbound communications, it was less economical than other methods in common use that also produced excellent results, so it was put on the shelf. Then came satellite communications which needed just such a low-noise receiver.

Visible to the naked eye, Echo could be seen in its spectacular race across the heavens, easily distinguished from stars of similar brightness as it streaked past them. Those who watched it were told that they had seen the prelude to a world-wide communication system.[3]

The first transmission of television pictures via an orbiting satellite was achieved on April 24, 1962, when pictures from the Massachusetts Institute of Technology laboratory at Camp Parks, California, were bounced off Echo 1 and received 2,700 miles away at Millstone Hill, Westford, Massachusetts. The pictures were described as "not clear" and of limited quality but recognizable.

THE NAVY'S MOON RELAY

If communications could be bounced off a balloon, why could not the moon or planets serve as reflectors of signals?

Pioneering in the use of celestial orbs as reflectors for radio relaying, the U.S. Navy developed a "communications moon relay (CMR)

[3] Echo 1 was launched on August 12, 1960. Echo 2, a 135-foot balloon satellite, was launched on January 25, 1964, from Vandenberg Air Force Base, California, into an orbit ranging from 642 to 816 miles high. With its skin described as "wrinkled like a prune" after 7½ years, one billion miles of flight, and 36,000 times around the globe, the 100-foot diameter sphere re-entered the earth's atmosphere in May 1968 and disappeared. It had traveled as low as 550 miles and as high as 900 miles above the earth.

system," noted as "the nation's first space telecommunications system." The project grew out of discoveries at the Naval Research Laboratory, where the feasibility of using moon reflection techniques for communications was demonstrated in 1951. The first public demonstration was conducted on January 28, 1960, between Maryland and Hawaii. Since then the system has been utilized for radio-teletypewriter, continuous wave (CW), voice, and facsimile traffic.

The transmitters are located at Annapolis and Opana; the receiving stations, at Cheltenham, Maryland, and Wahiawa, Hawaii, 5,000 miles apart. The moon must be above the horizon at both circuit terminals for the messages to get through. Therefore, the system can be used once a day from 3 to 12 hours depending on how far north is the path of the moon crossing the earth. The farther north the moon crosses the earth, the longer it is above the horizon and the longer the system can operate. The Navy radiomen have discovered that the moon is "no billiard ball" in smoothness, and its craggy surface necessitates special modulation technique for successful relay communications.

The U.S.S. "Oxford," a technical research ship, somewhere in the Atlantic on December 15, 1961, picked up messages transmitted from the U.S. Naval Research Laboratory Field Station, Stump Neck, Maryland, via the moon—the first messages sent from shore to ship from the lunar reflector. In 1962, the first two-way ship-to-shore moon-relay communication was conducted between Stump Neck and the "Oxford" at sea off South America, with transmission time to the moon and back approximately 2.5 seconds.

NAVY APPLIES A NEW SYSTEM

The Navy again maneuvered in space on July 22, 1963, when it orbited a satellite that for the first time utilized only forces of Nature to keep one face always pointed toward the earth. It was explained that a gravity-stabilizing system eliminates the need for weighty and complex devices such as electronically activated gas jets to hold the satellite stationary in its attitude toward the earth, just as gravity forces keep one side of the moon always facing the globe.

It was appraised as a promising development for communication satellites whose antennas must be pointed toward the earth in order to transmit and receive with utmost efficiency and low power. The ingenious satellite was designed for the Navy at the Applied Physics Laboratory, Johns Hopkins University, and, according to Dr. Richard B. Kershner, head of the laboratory's space division, "This system has been the dream of everyone in the satellite business."

A NATIONAL OBJECTIVE

A practical commercial communications satellite system became a national objective. The National Space Council recommended to President Kennedy that private companies jointly own the system while the government would establish broad technical and operating standards, also launch the satellites, and regulate the rates for space communications.

With this report as a guideline, the President, on July 24, 1961, set forth the Administration's policy favoring private ownership and operation of a communications satellite system capable of linking together "the farthest corners of the globe in the interest of world peace and closer brotherhood among peoples throughout the world." And he urged that it be developed for global benefit at the earliest practicable time.

Economic studies by industry revealed that despite the high initial costs, possibly $400 million or more, a communications satellite system eventually could become profitable.

The international communications business of American carriers is estimated at $135 million a year. Economic studies indicate that with a global satellite system in operation, it could be a billion-dollar-a-year business in the seventies.

By 1980, projections indicate there will be 500 million telephones in the world. And by that time, it is calculated that overseas messages to and from the United States will total about 100 million a year, requiring at least 10,000 overseas message circuits to handle traffic ranging from telephone and television to digital data, teletypewriter, and military services.

Raising their long-range sights to the turn of the century, when the world population is expected to reach five billion and telephones 1.5 billion, engineers foresee the need for about 200,000 circuits to handle the several billion messages that will flow between the United

States and overseas. Without satellites the engineers confess that they would be hard put to provide facilities to keep pace with the potential traffic demands.

To chart the way many problems must be solved. Since its formation in 1934, the Federal Communications Commission seldom if ever confronted a complex issue of such potential revolutionary impact upon world-wide communications, international relationships, and competition. To round up information upon which to base a decision, hearings were held to explore plans and procedures. Seven industrial organizations proposed systems: American Telephone & Telegraph Company, General Electric, General Telephone & Electronics, Hughes Aircraft, International Telephone & Telegraph Company, Lockheed Aircraft, and Radio Corporation of America. With variations, the majority favored a synchronous satellite system with three to six satellites in a 22,300-mile equatorial orbit.

To bring the matter to a focal point, a special industry committee, composed of representatives of ten United States companies engaged in international communications, was created by the FCC in July 1961 to work out the basis for a joint venture to own and operate a commercial satellite communications system. The committee recommended that the United States portion of a global system of communication satellites should be owned and operated, under government regulation, by the companies that the FCC might authorize to use satellites in providing communication services. This industry plan called for establishment of a non-profit corporation, owned by the companies engaged in international communications, to develop and operate the satellite system. As an alternative plan the Administration took under consideration the creation of a profitmaking publicly owned corporation.

SATELLITE CORPORATION FORMED

After a long debate centering on the question of government versus private ownership, the Senate passed the administration's communications satellite bill on August 17, 1962. Twenty days of floor debate and hundreds of hours of committee study were involved, and for the first time in 35 years the Senate's rules to break a filibuster were used before the bill was ready to go to President Kennedy, who signed it on August 31, 1962.

A joint government-industry organization—the Communications Satellite Corporation (Comsat) was created to establish and operate an international satellite communications system.[4] Cost of establishment was estimated at $200 million.

The President pointed out that many safeguards were provided in the statute to protect the public interest and that no single company or group would have the power to dominate the organization. On October 4, 1962, he named fourteen men to form the corporation under direct control of the government and to arrange for a public sale of stock; participating companies to own 50 per cent and the general public 50 per cent. Originally four United States companies: A.T.&T., ITT, RCA, and Western Union International participated in an investment of approximately $103 million.

Officially the orbiting "birds" known as "Intelsats" that do the relaying are owned by the International Telecommunications Satellite Consortium of 68 nations. Comsat, the United States representative, manages the consortium.

Ownership of the Intelsat system is based on use principle—a country invests in the system in proportion to its projected use of the satellite circuits. At the outset the proportion of United States ownership was 53 per cent, attributed to the fact that the majority of satellite communications either originated or terminated in this country. As the number of earth stations increases in other countries, it is expected that use of the total system by the United States will decline proportionally. Plans called for 70 ground stations in 40 nations by 1972; at the end of 1969, 43 stations were operating in 26 nations.

FOR PEACEFUL USE OF SPACE

"As we extend the rule of law on earth," said President Kennedy, addressing the United Nations on September 25, 1961, "so must we also extend it to man's new domain: outer space. . . . To this end, we shall urge proposals extending the United Nations Charter to the limits of man's exploration in the universe, reserving outer space for peaceful use, prohibiting weapons of mass destruction in space or on

[4] The Senate confirmed nominations on April 25, 1963; Leo D. Welch, chairman of the board and chief executive officer, and Joseph V. Charyk, president, of the corporation. James McCormack, Major General U.S. Air Force (retired) and a vice president of M.I.T., was chosen chairman, succeeding Welch on October 15, 1965.

celestial bodies, and opening the mysteries and benefits of space to every nation.

"We shall propose further cooperative effort between all the nations in weather prediction and eventually in weather control.

"We shall propose, finally, a global system of communications satellites linking the whole world in telegraph and telephone and radio and television. The day need not be far away when such a system will televise the proceedings of this body to every corner of the world for the benefit of peace."

Based upon President Kennedy's proposals, on November 27, 1961, the United States presented a plan to the United Nations Committee on the Peaceful Uses of Outer Space that would:

"Acknowledge that international law and the United Nations Charter 'extend to the outer limits of space explorations.'

"Have the United Nations Secretary General establish a central registry of space vehicles, maintaining a record of all space launchings and transmit data to members of the UN at their request.

"Use meteorological satellites to gain substantial progress in the atmosphere sciences and in weather analysis and forecasting and insure transmission of the satellite data to all countries through regional meteorological centers.

"Use communications satellites as orbiting relay points to form a global network available to all nations.

"Recognize the principle that outer space and celestial bodies are freely available for exploration and use by all nations and are not subject to claims of national sovereignty."

Ambassador Adlai Stevenson, United States delegate to the UN, on December 4, 1961, presented a resolution to the Political Committee of the General Assembly, calling for international cooperation in space to prevent "a runaway race into the unknown."

A forward step in that direction was taken on October 18, 1963, when the United Nations General Assembly ratified a 17-nation resolution, approved by the Assembly's Political Committee, to prohibit nuclear arms and other weapons of mass destruction in space.

And the United States and the Soviet Union agreed in the summer of 1963 to cooperate in space projects related to meteorology, communications satellites, and mapping of the earth's magnetic fields, experiments to begin in 1964. This was followed, in November 1963, by a general agreement between the two countries on a declaration of legal principles to govern the use and exploration of space. Based on the agreement, a draft submitted to the United Nations set forth that space and celestial bodies are not subject to national appropriation "by claim of sovereignty, by means of use or occupation, or by other means."

"The future of this country and the welfare of the free world depend upon our success in space," said Lyndon B. Johnson, then Vice President and head of the National Aeronautics and Space Council, addressing the American Rocket Society on October 13, 1961. "There is no room in this country for any but a fully cooperative, urgently motivated all-out effort toward space leadership. No one person, no one company, no one government agency has a monopoly on the competence, the missions, or the requirements for the space program. It is and must continue to be a national job. . . . We want to make the Space Age an age of peace. . . . There should be no obstacle to the establishment of international rules, preferably under the United Nations, for outer space."

And as President, in his first State of the Union message delivered to Congress on January 8, 1964, Mr. Johnson said, "We must assure our pre-eminence in the peaceful exploration of outer space, focusing on an expedition to the moon in this decade, in cooperation with other powers, if possible; alone, if necessary."

ELEMENTS OF A TREATY

While the United Nations in 1963 had declared outer space to be a province of all mankind that should be kept free from weapons of mass destruction, President Johnson on May 7, 1966, announced that the United States would seek an international treaty through the UN, to prevent any nation from claiming sovereignty to the moon and other heavenly bodies. He urged a treaty "to insure that exploration of the moon and other celestial bodies will be for peaceful purposes only." The essential elements of such a pact were outlined as follows:

The moon and other celestial bodies should be free for exploration and use by all countries.

No country should be permitted to advance a claim of sovereignty.

There should be freedom of scientific investigation, and all countries should cooperate in scientific activities relating to celestial bodies.

Studies should be made to avoid harmful contamination.

Astronauts of one country should give any necessary help to astronauts of any other country.

No country should be permitted to station weapons of mass destruction on a celestial body. Weapon tests and military maneuvers should be forbidden.

"I am convinced," added Mr. Johnson, "that we should do what we can—not only for our generation but for future generations—to see that serious political conflicts do not arise as a result of space activities. . . . We should not lose time."

Three weeks later the Soviet Union called for action by the UN General Assembly to restrict celestial bodies to peaceful exploration and scientific study for the benefit of mankind. Four points specified for inclusion in an international agreement or treaty, were similar to those enunciated by President Johnson.[5]

SPACE TECHNOLOGY

As the Space Age evolved it became more and more apparent that complete success in space depends not only on international legal agreements but upon international cooperation in electronic technology, which encompasses various categories interrelated through radio, radar, and television. All are boundless in scope: interspacial communication, space travel, satellite relays, guidance and control of space vehicles, telemetry, radio-radar astronomy, and scientific exploration.

Geophysics as a science began when man first used instruments to learn about the planet Earth. Now in the Space Age it becomes a study of the universe. And since electronics give it unlimited scope, new terms have been coined such as *geoscience electronics* and *geoengineering*—new professions applied to the earth, moon, stars, planets, and interstellar space.

[5] Sixty-two countries signed the treaty on January 27, 1967. It was ratified by the U.S. Senate on April 25, 1967. A space treaty on the rescue and return of astronauts was approved by the UN 28-nation Committee on Outer Space on December 16, 1967. It was signed by 44 nations on April 22, 1968. Signatories are bound to render all possible assistance to astronauts in the event of accident, distress, or emergency landing, and return them safely and promptly to representatives of the launching authority.

INTERPLANETARY COMMUNICATION

Interplanetary messages from another civilization have been predicted by the Jules Vernes in communications ever since the discovery of wireless. So many strange, unexplained noises have been heard from across the void of space that some have wondered if inhabitants of other planets were signaling among themselves or attempting to communicate with the earth. Skeptics, however, ascribe the tumult of hissing, spluttering sounds to atmospheric gibberish—static, cosmics, nuclear, or magnetic radiations—not to an intelligent source.

Similarly, if inhabitants of other worlds pick up radio signals and broadcasts from earth, the hash of noises might sound just as meaningless to them as splashes and clicks of static. Therefore, it is believed that television holds the greatest promise for communication with interspacial neighbors. They could more readily learn to understand pictures and diagrams and see evidence that the earth has abundant life. It would be a more complex riddle to decipher dots and dashes, or to comprehend spoken words in a tongue they had never heard.

To appreciate the significance of television in interplanetary communication, one has but to think how sensational it would be if suddenly some evening strange scenes and people of an alien civilization flickered phantom-like through space to appear on American screens. It could well become the greatest show on earth, and certainly more could be learned in an hour of pictures than in years of listening to code-like clicks that splash like atmospheric static. There seems no reason to believe, however, that another civilization would have the same TV system as the earth, or that their instruments could tune in terrestrial TV. To be sure, 70 years of radio have shed no clues of life elsewhere in the universe; so, it may be asked, Why anticipate that neighbors along the Milky Way will electronically break through space to show up on TV and override the Westerns?

Then, too, with radio and television traveling at the speed of sunlight, a habitable celestial orb might be so many light-years away that terrestrial radio broadcasts by the pioneer stations have yet to reach them, although they might be only 40 to 50 light-years distant.[6] The Harding-Cox election returns are still enroute, and they have yet to hear the voice of Coolidge, or a Roosevelt fireside chat. News of World War II is still lost in space, while television scenes of the World's Fair in 1939 have yet to arrive, provided, of course, that the pictures have not been bombarded by radiations that have ripped them asunder or distorted or completely erased them.

In any case there are plenty of targets out there. Within a distance of 500 light years—3,000 trillion miles—astronomers estimate there are at least 2,000 sun-like stars, and no one knows whether some are populated. If they are, scientists calculate that a powerful transmitter could be built to reach them. But it would take more than a lifetime to receive an answer; new generations, great-great grandsons would have to be alerted to listen.

Even among scientists doubt is found that such efforts would be "worth the candle." They contend, for example, that to bombard Mars with broadcasts is about as practical as a proposal made 100 years ago by Joseph Johann von Littrow of the Vienna Observatory that huge bonfires be lit on the Sahara Desert to signal the Martians.

Those who do not subscribe to the word "impossible" in extending communications deep into interstellar space base their hope for success upon the well-known fact that Hertzian waves travel into the infinite. Therefore, it is reasoned that if some other civilization is signaling, the means which might be used would be some form of radio operating in the electromagnetic medium. And, if inhabitants exist some 27 million miles away on Venus, Mars, or any other sphere, what system other than radio could man use to signal them?

Marconi was asked, when interviewed in New York in 1922, if he believed communication with Mars or other planets might someday be practical: "I would not rule out the potency of this, but there is no proof that signals come from another planet," he replied. "No one can say definitely that abnormal sounds on wireless originate on the earth, in solar eruptions, or in other worlds. We must investigate the matter more thoroughly before we venture a definite explanation.

"It is silly to say that other planets are uninhabited because they have no atmosphere or are so hot or so different from the earth. If there were no fish in the sea we would say life there is impossible. It

[6] Light-year: The distance covered in one year (about six trillion miles) by light traveling 186,000 miles per second.

is infeasible for man. Ask some of my material-minded friends, 'What is the practical advantage of communication with other planets; suppose communication is established?' I say the result would be the advancement of scientific knowledge by at least 200 years."

INTERSTELLAR POSSIBILITIES

Communications over interstellar, or interplanetary, distances seems like fantasy; so was intercontinental radio in 1900.

"It's not so important whether we just hear 'one, two, three,' or some other trivial message," said Professor Edward M. Purcell, Nobel Prize-winning physicist of Harvard, in a lecture at Brookhaven National Laboratory. "Just knowing we are not alone would change our entire philosophic outlook."

"With the new maser receivers being used in radio astronomy, reception of signals from 500 light-years away ought to be easy," said Professor Purcell. "I have argued that it is ridiculously difficult to travel even a few light years, and ridiculously easy to communicate over a few hundred light years."

Nevertheless, communication with interstellar neighbors, if any exist, is a project which he asserts has to be funded by the *century*, not by the fiscal year. He said it is not too soon, however, to have fun thinking about it and it is a much less childish subject to think about than astronautical space travel.

And referring to interstellar travel rather than interplanetary, he expressed belief that man is unlikely to meet any interstellar neighbors face to face. "The moral is that we aren't going anywhere, and neither are *they,*" he said. "Maybe you can get there by magic, but you can't get there by physics."

Leonardo Da Vinci, born in 1452, offered some prophecy along this line centuries ago. Observing a bird as an instrument working according to mathematical law, he declared it was within the capacity of man to reproduce such an instrument with all its movements.

"Observe," he said, "how the beating of the wings against the air suffices to bear up the weight of the eagle in highly rarefied air which borders on the fiery element." He conceived the idea, based upon study of birds in flight, of a flying machine and built a model to prove that man in flight was possible. And he predicted: ". . . He will cause himself to be carried through the air from East to West, and through all the uttermost parts of the universe. What is there which could not be brought to pass by a mechanism such as this? Almost nothing, except the escaping from death."

Communications, of course, is another thing. Radio, radar, and television beams can penetrate what scientists term "the crucial zone of life," and travel into the reaches of space, ignoring the untold dangers which confront a spaceship. Man in space must be shielded and protected from bombardment by powerful radiations and solar flares which he was not created to withstand in his earthly environment. The hazard of collision with meteoroids adds to the perils of exposure to swarms of hydrogen atoms, gamma, cosmic, and X-rays, nuclear radiations, and noxious gases that pervade space. And adding to the terror of it all, scientists warn that a single solar flare-up could snuff out a billion years of evolutionary advance! Searing heat, lethal climates, and other formidable forces of nature combine and conspire to create a smashing onslaught on the man who dares to go into the farflung interplanetary lanes, some of which pass through frigid zones estimated to be 400 degrees below zero.

ASTRO-ELECTRONICS

Radar's performance, together with radio echoes and mysterious noises from far beyond the orbit of the moon, inspired men of science to intensify their efforts in radio-electronics to gain supremacy in the international race for the conquest of space. They call this new field of science astro-electronics, operating in the vast laboratory of space —a boundless domain for expansion.

Little wonder that throughout the industrial world electronics is at the forefront in science, research, and engineering. Recognizing spatial guidance—from underwater to outer space—as a challenge of the day, International Business Machines Corporation points to the fact that in national defense and space exploration the electronic industry's assignment is to help develop and manage information and control systems for advanced weapons and space vehicles. Among the problems to be solved are:

Detection of enemy vehicles, from submarines to manned bombers and ballistic missiles, and the effective use of electronic countermeasures.

Guidance for deep space probes, and the relaying of information gathered by the vehicles to earth.

Surveillance of unknown satellites and calculation of their orbits.

Gathering intelligence data automatically, around the world and in space, and analyzing it for display in command control centers.

A solution to these problems may be found in a total information and control system, which gathers information from many remote points around the world, undersea, and in space. Data is communicated directly to computers at command control locations, and is processed immediately into meaningful form for display and decision, making it possible to apply electronic controls instantly where needed.

"The means for communicating information has never been so effective and far reaching," said Thomas J. Watson, Jr., chairman of IBM. "With radio and television, jet travel, countless libraries, and a highly articulate press, it is a remote and highly insulated individual who is not exposed to immense quantities of information. . . . The computer industry is finding the means to control this 'information explosion' and put its energies to work constructively just as scientists are learning to control nuclear energy. . . . The tremendous material progress made since World War I has been stimulated by our ability to communicate with one another."

WHERE TOMORROW BEGINS

"Tomorrow sometimes begins yesterday," IBM points out. And in a similar vein, the Boeing Company declares, "Tomorrow's business is largely based on technological studies being made today," for electronics is a major element in practically all missiles, spacecraft, commercial, and military aircraft.

Electronics guides the rockets, missiles, satellites, and space vehicles into orbit. Radar tracks them while the long arm of radio with its deft electronic fingers keeps them under control. Television and radio, powered by solar and atomic batteries, send back scientific data as well as pictures of the earth as seen from the satellitic altitudes.

Astroscientists call attention to the fact that guided missile history makes one point clear: The missile emerged through an evolutionary, rather than a revolutionary process. World War II rapidly accelerated the evolution as did the application of radioelectronics for communication and control.

Among the advances is a preset guidance system in which information relative to the course to be flown and the distance to be covered is fed into the missile's electronic computer, while other instruments within the "bird" sense deviations and automatically make corrections.

Then there is command guidance: The missile is radio-controlled on the basis of information the "bird" sends back relative to its position. A computer on the ground quickly compares the missile's position with its desired position at all times during the flight and makes corrections if it is off course by sending radio impulses to the missile's control system.

So accurate is all this that missiles hit targets thousands of miles down range. Satellites will not be lost in space, even in orbit around the sun, because electronics will keep an eye and an ear on them. In marked contrast, there was once a prewireless day when ocean liners left port not to be heard from until they reached the port of destination; not so with satellites a million miles away. Mammoth radio dish antennas scan the skies for them like a giant telescope. And these huge dish-like ears listen to the sounds from space and probe beyond the Milky Way, stimulating thought and imagination.

"Imagination rules the world," declared Napoleon. The modern version is that whoever breaks through in space is destined to rule the world and even the universe.

Electronic intelligence—Elint—is a new form of eavesdropping. Let a big shot blast off a launching pad anywhere in the world, and it is no secret at listening posts across the hemispheres. Indeed, there is no privacy or secrecy in the electronic game of hide-and-seek as played in the heavens. Space surveillance—Spa-Sur—is a new science that began in 1959.

Certainly Longfellow's poetic line, "I shot an arrow into the air, it fell to earth I know not where," does not apply to missilonics—the science of electronics related to missiles and satellites.

To shoot a rocket into the air without electronics would be a shot in the dark.

EXPANDING INDUSTRIES

It became clear that the first voyages into space had dispelled with dramatic suddenness any doubt about the future of science and industry in the realm of aerospace, defined as "the earth's envelope of air and the space above it."

Coined as a new word, "aerospace" came into the American language in 1958. Up to that time "air space" had been used. General Thomas D. White, then Air Force Chief of Staff, testifying before the House Astronautics and Space Committee on February 3, 1959, said: "Aerospace is a term which may be unfamiliar to some of you. I would like to define it. The Air Force has operated throughout its relatively short history in the atmosphere around the earth. Recent developments have allowed us to extend our operations further away from the earth, approaching the environment popularly known as space. Since there is no dividing line, no natural barrier separating these two areas, there can be no operational boundary between them. Thus air and space comprise a single continuous operational field in which the Air Force must continue to function. This area is aerospace."[7]

Aerospace industry sales in 1968 reached a record $30.1 billion, compared with $27.3 billion in 1967, according to the Aerospace Industries Association. The industry with 1.4 million workers is serviced by more than 50,000 suppliers in about every city and town in the country. It encompasses industrial America—plane makers, satellite and missile builders, electronic and communication companies, as well as manufacturers of chemicals, fuel, steel, aluminum, plastics, and paint; in fact, every activity engaged in all sorts of materials, products, and services so vital for exploration of the universe. The terrific thrust of aerospace to the nation's economy, to future expansion, and America's leadership is provided by research and engineering in which millions and millions of dollars are expended in what is heralded as a technological revolution.

The electronics industry in the United States passed the $10 billion sales peak for the first time in 1960, according to the Electronic Industries Association. Factory sales in 1960 totaled $10.677 billion; retail sales in consumer electronics increased from $3 billion in 1960 to $8.5 billion in 1968. By 1969 the electronic industry's over-all total was more than $24 billion. Looking ahead, projections indicate

[7] Air space is considered to be the space between the earth at sea level and an altitude of 50 miles; inner space, from the earth to the 500-mile altitude; and from there on, outer space.

that retail sales in consumer electronics could exceed $12 billion by 1975.

Growth factors include all phases of color TV, radio, phonographs, and tape recorders. Other areas of vast potential feature expanding use of industrial control and processing equipment as well as computing and data processing installations. Expanding applications are foreseen for semiconductors, integrated circuits, micro-electronics, electronic devices in the fields of education and medicine as well as in aviation and astronautics. Extension of electronics throughout communications looms unlimited together with advances in industrial, military, facsimile, data-phones, picturephones, microwaves, laser and nuclear technologies.

IN THE FIRST DECADE

In the first six years of the satellitic era, the United States successfully launched several hundred satellites and interplanetary probes. Activity stepped up during 1963 when America was reported to have put at least 55 scientific and military payloads into orbit. By spring of 1964, more than 230 man-made objects were believed to be orbiting in space, representing what remained of the 575 devices the United States and the Soviet Union had launched since October 1957.

By 1970 more than 4,225 man-made objects had been hurled into space since the first Sputnik; 1,826 were still adrift, according to the Air Force Defense Command. The debris consists mostly of inoperative satellites, nose cones, burned-out rocket castings and metal fragments. Some of the litter eventually will be consumed at re-entry into the earth's atmosphere, while the larger items continue in orbit for hundreds if not thousands of years.

A huge radar unveiled at Eglin Air Force Base, Florida, in 1967 tracks 95 per cent of the objects as far out as 2,650 miles. Differing from regular radar the installation has a sloped face minus the rotating antenna and is housed in a building more than a block long and 15 stories high. Instead of scanning the skies mechanically it scans electronically. As many as 5,184 beams of radio energy are projected at space targets and the energy bounces back.

It is pointed out that any object orbiting the earth crosses the equator twice a day, and since the Eglin radar faces south it can detect, track, identify, and predict the orbits of many objects. Three computers linked with the radar can determine in a few minutes whether an object is old or new, reveal its orbital characteristics, and indicate where it will land if on a ballistic trajectory, or a satellite about to be gripped by the earth's gravity. Conceived in 1961 the system was developed for the Air Force by the Bendix Corporation at a cost of approximately $62 million.

During the first decade in space (1957–1967) the United States orbited 500 satellites, accomplished 13 successful moon missions, and logged 1,994 hours of manned flight. The Soviet Union launched 250 satellites, had 533 hours of manned flight, and was credited with 8 lunar successes.

President Johnson's report to Congress on aeronautics and space activities in 1968 noted 64 space vehicles launched by the United States, 74 by Russia. The total for 11 years through 1968 was 606 American launchings; 358 Russian. Total United States spending for space activities up to 1969 was reported at $56,727,500,000, which included $17 billion by the Defense Department and $38 billion by NASA; the remainder by other agencies, such as the Atomic Energy Commission.

"Since the greatness of a nation depends to an important degree upon mastering and putting to use technological advances," said the President's report, "new goals and new directions must be established in space to maximize the existing and potential benefits of this new national asset."

Sir James Jeans once asked, "Is our importance measured solely by the fractions of space and time we occupy—space infinitely less than a speck of dust in a vast city, and time less than one tick of a clock which has endured for ages and will tick on for ages yet to come? Let us reflect that mankind is at the very beginning of its existence; on the astronomical scale it has lived only for a few brief moments, and has only just begun to notice the cosmos outside itself. It is, perhaps, hardly likely to interpret its surroundings aright in the first few moments its eyes are open."

X   Communications Satellites

> The electronics men have just gone
> beyond my comprehension.
> —*Herbert Hoover*

A NEW ERA OPENED IN LONG DISTANCE COMMUNICATION WHEN Telstar, a communications satellite developed by the American Telephone & Telegraph Company, rocketed into orbit on July 10, 1962.

As the 170-pound sphere whirled around the globe on its sixth orbit over the Atlantic (apogee 3,531 miles, perigee 593 miles)[1] memorable records were established. The American flag waved on TV screens at home and abroad while the sound track played "The Star Spangled Banner" and "America the Beautiful." From Andover, Maine, Frederick R. Kappel, chairman of the board of A.T.&T., telephoned to Washington via the satellite to demonstrate the new voiceway. And in France television reception was so clear that operators at the Pleumeur-Bodou station in Britanny reported the quality of transmission was equal to that of a station 20 miles away. In Britain the picture was seen to waver as Telstar dipped toward the horizon.

The next night France added luster to the international TV performance by relaying a video tape to millions of viewers in the United States. Communications Minister Jacques Marette was seen as he introduced the program featuring Yves Montand singing "La Chan-

---

[1] Apogee is that point in the orbit farthest from the earth; perigee the point nearest the earth.

sonnette," Michele Arnaud singing "Deux Enfants au Soleil," and Michel Aubert, guitarist, playing "Chanson de l'Été.

About three hours later, as the sapphire-studded Telstar made its next orbital pass, the British station at Goonhilly in Cornwall put on several slides and test patterns but made no attempt to entertain. The Telstar pictures photographed off TV screens were printed on the front pages of American newspapers.

The nerve center of activity in the United States was Andover, Maine, near which a 210-foot sphere of inflated silvery fabric housed a huge 380-ton horn-like antenna that could rotate and twist as a mammoth "ear" toward any part of the sky. This mountain-ringed 1,000-acre site among the pine trees was well named "Space Hill," while signs along the roadway pointed the way to the "Earth Station for Communicating by Satellite"—a $10 million facility, 1,000 feet above sea level.

The huge spherical cover over the antenna, known as a radome, is supported by air pressure. It is 210-feet wide and 161-feet high, enclosing about twice the space of the seating area in New York's Radio City Music Hall. The radome's skin of synthetic rubber and fabric, "transparent" to radio waves, weighs about 20 tons and if laid out flat would cover three acres.

Telstar's radio transmitter, rated at 2¼ watts, was of feeble power. Nevertheless, Andover's "big ear," 177 feet long and 94 feet high, was designed to hear the faintest electromagnetic whispers. The incoming impulses were greatly amplified at the nearby control center, giving the TV pictures, voices, and coded data plenty of strength to travel over telephone lines and to reach the television networks.

Described as "a switchboard in the sky," Telstar was more than that. It received signals from the ground, amplified them billions of times, and relayed them on another frequency. For instance, signals flashed to it on 6,390 megacycles were rebroadcast on 4,170 megacycles.

The satellite, 34.5 inches in diameter, was a wonder of the Electronic Age. It had 72 flat facets, and mounted on 60 of them were 3,600 solar cells to convert light from the sun into electrical current. Man-made sapphire coated the cells for protection against radiation bombardments. Operating power generated by the solar cells was only 15 watts, not sufficient to keep the apparatus functioning all the time. For that reason the satellite was equipped with a command-obeying system, which threw electronic switches in response to coded radio signals from the earth. When the circuits were not needed they were turned off to conserve power and give the solar cells a chance to recharge the satellite's storage battery.

A satellite such as Telstar acts as an antenna tower in the sky. In effect it takes the place of a relay tower, which would have to be 475 miles high over the mid-Atlantic, or of ten transatlantic TV cables, which would cost a quarter of a billion dollars. Carrying more than 15,000 electronic parts it raced along its orbit at a top speed of 16,000 miles an hour. It was equipped with 1,064 transistors and 1,464 diodes. Fifty million dollars were spent to design and build it, and $2.7 million was paid the government for launching it—proof that private enterprise and government can work in harmony in bringing new marvels of science and industry into service. About 1,250 different companies—the majority small businesses—participated in the project as subcontractors and suppliers.

It was emphasized that Telstar was not a single invention, but was based on at least 16 patents obtained by the Bell Telephone Laboratories over a quarter of a century, and oddly enough none was made specifically for space purposes. George C. Southworth of the research staff discovered that radio waves could be guided through hollow pipes, and in 1936 he patented his waveguide system. At the time it was looked upon as an interesting experiment, and little practical use was foreseen. Southworth predicted that the technic "beyond this point is a matter for the future." He was right. Waveguides became important in Telstar and at the ground stations as well as in radar.

The giant trumpet-shaped antenna used as the earthly ear was patented by Alfred C. Beck and Harald T. Friis in 1947. Also out of the past came the transistors, solar cells, and many of the other devices necessary to create satellite communications. The helical antenna atop Telstar was patented in 1946 by Harold A. Wheeler of Wheeler Laboratories, Inc. But it was not until an antenna capable of circular polarization in a doughnut pattern was needed for the satellite that a use was found for it—15 years later!

Acclaimed in front-page headlines, Telstar inspired more than 1,400 editorials in newspapers during its first month of operation;

hundreds of feature stories and magazine articles told of its triumph.

The first formal exchange of TV programs—panoramic views of cities, landmarks, and a variety of activities—made July 24, 1962, a historic date in the annals of television. Europeans caught a glimpse of Niagara Falls, the Statue of Liberty, a baseball game played by the Phillies and Cubs at Chicago, and other American scenes.

Americans were shown Europe's cultural diversity—the entrance to the Louvre in Paris, the Colosseum in Rome, the Tower Bridge over the Thames in London, and scenes from Stockholm, Belgrade, and other cities. Estimates put the 18-nation audience at 100 million persons.

President Kennedy said that the achievement "while only a prelude already throws open to us the vision of an era of international communications." He added: "There is no more important field at the present time than communications and we must grasp the advantages presented to us by the communications satellite to use this medium wisely and effectively to insure greater understanding among the peoples of the world."

Asked what he thought about the Telstar broadcasts, former President Herbert Hoover said, "The electronics men have just gone beyond my comprehension. I belong to a generation that just doesn't grasp all that."

TELSTAR MILESTONES

Aside from its spectacular TV performance Telstar made and reported 112 important measurements and observations each minute, providing information of value to the future of space communications, all of which was correlated by electronic computers.

Among other milestones in communications established by Telstar during 1962 were:

July 12: First transatlantic telephone conversation via satellite was held by American and British technicians; Eugene McNeely, president of A.T.&T., talked with Jacques Marette, French Minister of Communications, the next day.

July 17: First color TV via satellite was relayed across the Atlantic by British engineers from Goonhilly station in Cornwall to Andover, Maine.

July 31: Former President Dwight D. Eisenhower being greeted by Premier Tage Erlander of Sweden was seen via satellite relay from Stockholm; Eurovision sent the program to England, from where it was relayed via the satellite to the United States.

August 18: Seven pages of New York newspapers were transmitted by facsimile to and from Telstar.

August 22: International edition of *The New York Times* carried several articles transmitted from New York to Paris via Telstar; 5,000 words were sent at a rate of 1,000 words a minute, 16 times as fast as other transatlantic copy sent by way of conventional radio and cable.

August 27: Telstar used to synchronize master clocks in the Royal Greenwich Observatory at Herstmonceux, England, and the Naval Observatory in Washington, D.C., with an accuracy of ten millionths of a second compared to conventional radio measurements of about one or two thousandths of a second.

October 3: Within 40 minutes after Commander Walter M. Schirra, Jr., rocketed into orbit, taped pictures of the blast-off were Telstared to Europe where many millions looked in.

October 11: Opening of the Ecumenical Council in Vatican City, assemblage in St. Peter's Square, were relayed to American TV screens.

Something went wrong up there in the sky on November 23, 1962. Telstar stopped functioning as a relay station. It continued in orbit with the telemetry equipment still furnishing information on the amount of radiation encountered, temperature within the sphere, and effects of radiation on the transistors and solar cells. For five months it had been used in more than 250 tests and in 400 demonstrations of its capabilities in communications; it had relayed 47 transatlantic telecasts.

Suddenly on January 3, 1963, Telstar came to life after six weeks of "rest." High-level radiation created by a powerful nuclear test bomb exploded by the United States over the Pacific was blamed for bathing the transistors with ions (electrical charged particles) that prevented them from working properly. Engineers simulated the effect in the laboratory and traced the ailment to a single transistor called the "zero gate," designed to react to short pulses in command signals. They learned how to dissipate the ions clustered around the transistor, and once again Telstar obeyed commands, and TV relays resumed. A new long-distance record in radio servicing had been established. Nevertheless, on February 21, 1963, Telstar went mute again, but it is expected to remain in orbit for about 200 years.[2]

As *The Times* pointed out, the brilliant initial successes underlined urgency of the need for a space law because "it represents the first step toward integrating this new dimension into our day-to-day lives."

MAINE AS A SPACE CENTER

Satellites would orbit without earthly contact were it not for ground stations to signal them and pick up what they relay.[3] Andover, Maine, having proved successful as the Bell System's satellite communication center, as evidenced by Telstar successes, it may be expected that "as Maine goes" in radio, so it will go forward in interspatial communications. The Pine Tree State as early as the First World War, proved to be an ideal site for clear radio reception, and for that reason the U.S. Navy located one of its most important stations, NBD, at Otter Cliffs, east of Bar Harbor on Mount Desert Island.[4] Static and stray radio interference were at a minimum, blocked off by the mountains, so that reception from Europe and ships at sea was unusually dependable. On one occasion an SOS was intercepted from a British ship in distress 900 miles east of Bermuda, but was unheard by any other station along the eastern seaboard from the St. Lawrence to the Amazon. Therefore, based upon Maine's historic record in communications, even the faintest signal from a

[2] Telstar 2, launched on May 7, 1963, into an orbit ranging from 600 to 6,700 miles above the earth, was basically the same as Telstar 1 except that transistors and command circuits were designed to be more resistant to radiation. Communication to the satellite was on 6,390 megacycles, and relays from the satellite were on 4,170 megacycles—the same frequencies used by Telstar 1. On July 16, 1963, Telstar 2 ceased to function, but during its 622nd orbit on August 12, operation resumed and it went on to establish new records, including relay to Europe of the Johnson-Humphrey inaugural on January 20, 1965, and the funeral of Sir Winston Churchill telstared to the United States on January 30, 1965.

[3] Other ground stations for satellite communications are located in Great Britain, France, Italy, Germany, Japan, and Canada.

[4] The U.S. Naval Security Group Activity is located at Winter Harbor, Maine. The U.S. Navy low-frequency, 2-million-watt station NAA at Cutler, Maine, is rated as the most powerful in the world.

celestial station is not likely to go unheard. There are more than 75 million telephones in the United States, and it is estimated that the count will be 235 million by 1980. If only a fraction of them ping voices off the satellites, space will pulse with conversation.

MILITARY SATELLITES

With demands for international communications growing so fast, cables, radio, satellites, and beams of light will all be needed to serve the future.

A military communications satellite system planned by the Defense Department specified 24 to 30 satellites to link United States strategic forces around the world. A number of the "stations" were orbited from 1965 on, so that the complete system might be in operation in the 1970s.

Adhering to the schedule an Air Force Titan 3-C rocket put eight satellites into orbit above the equator on June 16, 1966. Seven were for communications and one for research. Traveling in paths about 21,000 miles above the earth, the 100-pound relay stations were to serve as communication links between Washington and United States military forces around the world. In addition two 220-pound research satellites OV-9 and OV-10 were polar-orbited on December 11, 1966.

To double the Defense Department's satellite communications network, on January 18, 1967, the Air Force entrusted eight additional relay satellites to a Titan, which dropped them into slightly different paths to provide a full-time link with American forces in Vietnam and elsewhere. While the original seven military "switchboard" satellites were in range for relaying to any specific ground terminal about 80 to 90 per cent of the time the supplemental eight were expected to fill in any gaps. Each satellite designed to link ground stations 10,000 miles apart was built to operate for at least 1½ years powered by solar cells.

Six military satellites boosted into high-altitude outposts by a Titan-C on July 1, 1967, were placed one-by-one in separate orbits ranging from 20,652 to 20,885 miles high. Four of the spacecraft were radio relay stations that fitted into a network of 15 already in operation. The fifth was assigned to test a new communications system for frontline troops. The sixth, a 437-pound craft, was designed to sprout ten booms up to 150 feet long to determine if the slight difference in gravity's pull on them would stabilize satellites in orbit. No. 6 also carried two TV cameras to televise the earth in color.

In an 8-in-1 shot by a Titan rocket on June 13, 1968, the Air Force orbited eight communication spacecraft aimed to increase the Defense Department's "switchboard-in-the-sky" to 26 satellites for contact with military bases. And on August 6, 1968, a secret "experimental payload" launched by the Air Force was listed as the first classified flight from Cape Kennedy since 1963. That was followed on September 26, 1968, by the orbiting of four satellites including one built by the Massachusetts Institute of Technology to conduct experimental tactical radio tests of jam-resistant messages to strategic units on land, sea, and in the air.

As part of an Air Force sponsored project in space communications, a number of Lincoln Experimental Satellites (LES), designed and built by the M.I.T. Lincoln Laboratory, were orbited. LES-5 lofted in July 1967 was rated as the first all-solid-state UHF-band communications satellite.

A giant 1,600-pound experimental Tacomsat (Tactical Communications Satellite) launched by the Defense Department from Cape Kennedy on February 9, 1969, was foreseen as a possible forerunner of a military communications network of three or four orbiting transmitters. Cylindrical in shape, 8-feet wide and two stories high, it was "parked" 22,300 miles above the Pacific off the coast of South America. Built by Hughes Aircraft Company this first Tacomsat was designed to transmit several thousand two-way telephone calls at one time. Its operational life was estimated to be at least three years, during which time it would be capable of radiating signals for reception by all types of ground stations, including small antennas carried on the backs of infantrymen in combat.

A joint project by the armed forces introduced in 1969 as Tacsatcom (Tactical Satellite Communications) beamed signals through a Tacsat satellite hovering in synchronous orbit above the mid-United States. Experimental super-high frequency (SHF) sets demonstrated two-way voice communications between Washington and military outposts around the country. Five versions of the receivers featured: a one-man pack device for listening only (alert receiver); a team pack that can be transported by three men; a Jeep-mounted unit; a shelter unit transportable by truck, aircraft, or helicopter; and an airborne version. Except for the "listen only" one-man outfit, they are capable of two-way voice communications. The Jeep, shelter, and airborne installations can handle printed messages as well.

VELA "WATCHDOGS"

As the first in a series known as Vela, the United States on October 16, 1963, sent up two patrolling satellites designed to detect any violations of the nuclear test-ban treaty. Both of these "watchdogs" were maneuvered into circular orbits about 57,000 miles above the earth. So that they might differentiate between natural radiation and clandestine nuclear explosions, the two satellites would always be at opposite sides of the earth.

Velas 3 and 4 were launched on July 17, 1964. Numbers 5 and 6, 20-sided satellites with most of their surface covered with solar cells, were sent into 60,000-mile-high orbits on July 20, 1965. They measured 54 inches in diameter, weighed 334 pounds and carried 94 pounds of instruments. Capable of detecting blasts far out in space they also could "look" into the atmosphere to record nuclear explosions near the earth's surface.

Two 731-pound Vela spacecraft were included in a piggy-back line-up atop a Titan 3-C booster when the Air Force launched five satellites simultaneously on April 28, 1967. The Velas engineered to detect sneak nuclear tests were hurled into an elliptical orbit from 5,400 to 69,000 miles above the earth. The other three vehicles were for scientific exploration, such as monitoring radiation and measurement of X-ray outbursts from the sun. A duplicate quintuplet launch by the Air Force on May 23, 1969, featured twin 755-pound Velas; the other three vehicles were to check various types of radiation.

Numerous secret "spy-in-the-sky" unpublicized vehicles have been sent into space. Estimates in 1969 indicated that more than 400 military satellites had been launched in a decade by the United States and the Soviet Union. Some of them were described as "highly sophisticated advances in automatic techniques for unmanned satellite systems." One American version "IS" (Integrated Satellite), first launched in 1968, was reported to carry television as well as high-resolution photographic equipment and instruments to detect infrared radiation which could reveal underground installations such as factories or missile launching sites.

PARTNERS IN COMMUNICATIONS

While satellites as newcomers in communications gained news headlines, undersea links were being modernized by electronic technology. Coaxial cables, semiconductors, new deep-sea repeaters (amplifiers), and other devices led to extension of overseas voiceways that can handle 720 simultaneous conversations as well as radio and television.

Transocean telephone conversations totaled 15,200,000 in 1968, according to the A.T.&T. Estimates indicate approximately 18 million in 1969 and tens of millions by 1980.

An illustration of how satellites, land-lines, and cables supplement each other is found in the partnership exhibited when two land-line links to Mexico City brought the 1968 Olympics to the TV audience in the United States. The terrestrial lines also connected to satellite earth stations near San Francisco and Etam, West Virginia, for relay to Hawaii, the Orient, and to Europe.

Indicative of continued advances the A.T.&T. 1968 annual report stated, "Service over satellite circuits to Chile and Panama began last summer and fall. We are now using some 1,100 cable and more than 400 satellite circuits (as well as shortwave radio) for overseas service and we expect to use many more of both. Satellite circuits are leased from Comsat, of which we are the largest customer. Service is available to 210 countries and territories—including service to Vietnam."

RELAY

While cables were being submerged in the oceans, new satellites were being planned and orbited. Relay, a 172-pound communications satellite designed and built by RCA for NASA was launched on December 14, 1962, into an orbit from 800 to 4,000 miles high.

Plagued from the outset by abnormal power drain, the satellite was diagnosed as "too sick" to relay an 11-nation Christmas pageant that was to demonstrate its versatile capabilities. Once a faulty transponder (transmitter-receiver) was isolated and a substitute shifted into operation, power was restored, and Relay on its 169th orbit went into action on January 3, 1963.

Six days later TV filmed ceremonies of the first public showing of Leonardo da Vinci's "Mona Lisa" at the National Gallery of Art in Washington were relayed to Europe with excellent results. And following that success hundreds of engineering tests, including color television, were conducted with foreign participants operating ground stations in England, France, Italy, and Brazil.

When President Kennedy, acting on behalf of the United States Government, on April 9, 1963 conferred honorary United States citizenship on Sir Winston Churchill, the veteran statesman watched the Washington ceremony on television in London, transmitted via Relay to Great Britain and the European continent.

To test the feasibility of quick international diagnosis, brain wave data graphs (electroencephalograms) were flashed across the Atlantic on April 25, 1963, from the Burden Neurological Institute in Bristol, England, to a meeting of the National Academy of Neurology in Minneapolis, Minnesota, and a diagnosis was sent back within minutes via Relay.[5] As a further demonstration of the satellite's versatility, a color illustration was sent to London and back to New York, where color separation plates were made for the cover of *Electronics* magazine.

Acclaimed as a durable and powerful satellite, Relay, as well as Telstar, was used extensively by the TV networks to cover a European trip by President Kennedy in July 1963, the death of Pope John XXIII, and the election as well as the coronation of Pope Paul VI.

Relay was equipped with an electrochemical timer—an electrolytic material that would eat away a connection in the power supply line at a predetermined date. It was calculated that by the end of a year the mission would have been accomplished; and therefore, as an engineer explained, it was deemed good sense to silence it so that its signal would not be added to the already limited radio-frequency spectrum.

But Relay went chattering along into 1964. Engineers speculated that failure to shut off was caused by space temperatures much colder than anticipated, so that erosion was retarded. Consequently, Relay,

[5] Electrocardiograms telstared across the Atlantic on June 11, 1963 were reported clear enough to be meaningful to physicians in Paris.

having traveled several hundred million miles, continued on its way, adding to its record of more than 2,000 successful operations, including about 300 hours of intercontinental television and 630 demonstrations in transmission of voice, facsimile, data processing, and other signals. It even passed through the Van Allen radiation belt with no ill effects. Its series of firsts via space included: a space link to South America, network color TV, computer typesetting, two-way transpacific TV between the United States and Japan, satellite-relayed telephone conversations between America and West Germany, and many news events.

Relay 2, launched on January 21, 1964, was modified to enhance reliability and performance, but its basic structure and mission were identical to Relay 1—to provide broad-band communications. No shut-off timer was on board this time for silence on a predetermined date, so that it could continue operation indefinitely to test the utility and longevity of a communications satellite traveling on an orbit ranging from 1,325 to 4,600 miles high.

SYNCOM

To test a high-altitude synchronous communications system, Syncom, an 86-pound satellite, was launched on February 14, 1963. Developed by the Hughes Aircraft Company under a contract from NASA, Project Syncom was planned as the first attempt to orbit a communications satellite 22,300 miles above the earth—and "park" it there. At that altitude Syncom would travel at the same speed as the earth in rotation and would appear to be suspended in space, although actually traveling 6,875 miles an hour. From the vantage, above the mouth of the Amazon River, Syncom could "see" 40 per cent of the earth. It was calculated that three Syncoms spaced at equal intervals could form a vast telephone and TV network circling the globe to provide continuous, uninterrupted service for most of the world.

Something went wrong shortly after the launch; radio and telemetry contact failed, and the satellite, reported to have cost $11 million, was lost in disappointing silence somewhere between Madagascar and Africa's east coast. Engineers speculated that the spinning axis might have been tilted away from the ground stations, or that the satellite might have been nudged off course or even destroyed by a malfunction of the kick rocket.

SYNCOM 2

A second Syncom was boosted into orbit on July 26, 1963, while hopes of space scientists ran high that it would retrieve the "lost glory" of Syncom 1. Two ground stations established for the experiment were located at Lakehurst, New Jersey, and aboard the U.S.S. "Kingsport" in the harbor of Lagos, Nigeria.

Five hours and 33 minutes after launching, when the drum-shaped 79-pound satellite soared over Africa at a high point of 22,548 miles, a small motor, or timer, arrested the vehicle as it drifted toward the planned synchronous orbit 22,300 miles above the earth. Operators in control reported all systems functioning extremely well as the satellite received and transmitted a recording of "The Star-Spangled Banner" back to the "Kingsport"; a voice tape and teletype data transmitted from a station near Paso Robles, California, were relayed over 7,700 miles.

For three weeks the satellite drifted toward the position where it would be "on station." From time to time its speed and course were corrected by ground-controlled jets with such efficiency that the intricate maneuvering was rated as "one of the outstanding feats in the history of space flight." When it reached the desired location above the equator at the mouth of the Amazon river, radio signals from the ground fired its nitrogen-jet brakes to halt the flight at that point and orient the antenna. Then the orbital speed was increased somewhat by the use of solid-fuel altitude control jets so that the timing of the functional orbit would synchronize with the speed of the earth's rotation; thus the satellite would appear to "stand still" over the same area of the globe. It orbited in a figure-eight pattern along the meridian from Brazil up the mid-Atlantic from latitude 33 degrees south to latitude 33 degrees north.

Once "on station" ready for action, Syncom demonstrated its capabilities in various tests, including TV experiments. And on August 23, 1963, President Kennedy and Prime Minister Balewa of Nigeria exchanged telephone greetings via Syncom, acclaimed as "a technical tour de force."

Three months later—November 22—the satellites he so vigorously encouraged relayed tragic news: An assassin's bullet killed President Kennedy at Dallas, Texas. For 3½ somber days electronic communications performed as at no time in the decades of broadcasting.

President Johnson, who took over after Kennedy's death, was seen across the hemispheres as he addressed a joint session of Congress and the nation, resolving to "Let us continue" what John F. Kennedy had resolved "to begin" on his Inauguration Day. The leader who had exclaimed, "Let us explore the stars," was gone. He had been determined to make the United States the leading spacefaring nation on the empyreal ocean. It was he who called upon America to embark on "a great new enterprise," and to take a leading role in space achievement to gain a position of pre-eminence, to open the benefits of space to every nation and to reserve outer space for peaceful purposes.

THE ENTERPRISE CONTINUES

A new communications satellite Syncom 3 was tossed onto the empyreal ocean on August 20, 1964. Maneuvered into orbit 22,300 miles above the Pacific, it relayed the Olympic games from Tokyo, October 10–24. And as an exchange feature the World Series on TV was seen in the Far East for the first time.

An Early Bird (Intelsat 1), the first commercial communications satellite, was launched for Comsat on April 6, 1965. Weighing 85 pounds the hat-box-shaped "switchboard" moved along a 22,300-mile-high orbit above the Atlantic at the equator traveling at the same speed as the earth's rotation. It was designed by the Hughes Aircraft Company to handle 240 simultaneous telephone conversations as well as television. To complete a world-wide system Comsat ordered four additional satellites in October 1965 at a cost of about $11,730,000. Three of the "birds" would be stationed over the Atlantic, Pacific, and Indian oceans.

On May 2, 1965, having achieved proper position over the equator, Early Bird demonstrated its usefulness in an international TV program. Millions of spectators in North America and Europe looked in on an exchange of scenes as TV skipped from the United States, Mexico, and Canada to England, France, Spain, Switzerland,

Sweden, and Italy. Physicians in Geneva watched a heart operation performed by Dr. Michael E. DeBakey in Houston, Texas. The pictures were of excellent quality and the close-up of the aorta valve operation was described as "almost uncanny." All in all it was a vivid one-hour demonstration of international television's educational, entertainment, and news potential.

As evidence of the police power of satellcasting, pictures of several persons wanted by the FBI and Scotland Yard were sent far and wide via Early Bird. The photograph of one culprit wanted in connection with a Montreal bank robbery in 1961 was recognized by a boat repairman at Fort Lauderdale, Florida, where he was arrested.

FIRST COMMERCIAL SERVICE

The first commercial telephone service via satellite was inaugurated by President Johnson on June 28, 1965, with Early Bird as the sky switchboard. Officials of six nations, including Prime Minister Wilson of Britain and Ludwig Erhard, Chancellor of West Germany, participated in opening the Comsat phone system that added 18 new telephone circuits to the existing 107 linking the United States and Europe and 6 to the 34 with Canada.

Said the President: "Other satellites in days to come will open communications pathways for the world. But we are especially pleased that this first service brings closer together lands and people who share not only a common heritage but a common destiny—and a common determination to preserve peace, to uphold freedom, to achieve together a just and decent society for all mankind. In these times, the choice of mankind is a very clear choice between cooperation and catastrophe. Cooperation begins in the better understanding that better communications bring."

As an outstanding illustration, just as aviation brought African nationalist leaders into direct contact, similarly satellite communications was foreseen contributing to growth and understanding among African nations. The Space Age held promise of opening up the Dark Continent through formidable economic, social, political, and technical advances long deterred because of limited communications between African nations as well as the outside world. With the advent of satellites any nation with a ground station could directly telephone or radio to Europe, the United States, India, or South America, and conduct intra-continental communications including computerized data.

Nicknamed Lani Bird, a communications satellite, Intelsat 2, reported to have cost more than $2 million to build and $3.7 million to launch, was orbited on October 26, 1966, as a Comsat link to Asia.

Failure to achieve the proper path put it in an elliptical orbit ranging from 2,127 to 23,000 miles above the Pacific, whereas a circular 22,300-mile route would have been ideal. Fortunately, however, the craft was maneuvered so that its 240 channels could be used eight to nine hours daily for telephone and television between Hawaii and the United States; more than seven hours daily between this country and Japan, and about four hours with Australia.

To demonstrate its capability the Notre Dame–Michigan State football game on November 19, 1966, was sent via land line from East Lansing, Michigan, to the Comsat station at Brewster Flat, Washington, which relayed the color TV pictures via Lani Bird to the terminal station at Paumalu, Hawaii. It marked the first TV between the mainland and the island, and during the half-time period of the game, scenes from Waikiki Beach were telecast throughout the United States.

Lani Bird 2 (a second Intelsat 2 also called Pacific 1), a 192-pound craft launched on January 11, 1967, went into a circular orbit 22,300 miles above the Pacific to beam telephone, teletype data, radio, and television between North America and the Far East. It was supplemented on September 27, 1967, by Pacific 2, a $2.7 million, 192-pound, drum-shaped satellite "parked" in orbit above the central Pacific.

Comsat's fourth communications satellite (a third Intelsat 2) was stationed 22,300 miles above the Atlantic on March 22, 1967, to serve as a "switchboard" between the United States, South America, and Africa. It was nicknamed "Canary Bird" in recognition of the ground terminals on the Canary Islands.

An electrical malfunction in the Thor Delta's guidance system led to destruction of the rocket and Intelsat 3 by the range safety officer 108 seconds after the launch on September 18, 1968. Had it achieved the intended orbit over the Atlantic off the coast of Brazil, it would have more than doubled the world's capacity to relay telephone and television transmissions.

To achieve that goal another 642-pound Intelsat 3-A was sent up on December 18, 1968. By that time three new ground stations were ready for operation at Etam, West Virginia; Cayey, Puerto Rico; and Tulancingo, Mexico, increasing the total operational earth terminals to 22 stations in 14 countries. By 1973 the network is expected to include 70 stations in 62 countries.

Engineered to relay 1,200 telephone calls or beam four color telecasts, the third in the Intelsat 3 series was lofted over the Pacific on February 5, 1969. The $6-million satellite weighing 632 pounds was orbited to expand transpacific service compared to Intelsat 2, which could handle 240 telephone circuits or one telecast. Electrical problems, however, led to replacement of Intelsat 3 in May 1969 when another $6.8 million payload also designated as "3" was stationed over the Pacific. Another launched on July 25, 1969, went into the wrong orbit when the upper stage of its rocket misfired. The sixth Intelsat 3 was scheduled for 1970 at the opening of which three Intelsats were in orbit.

MOLNIYA SATELLITES

While the Intelsat system was evolving, the Soviet Union sought "to verify the possibility of organizing a (domestic) communications system with the use of several satellites." First in the series, Molniya 1, pronounced mol-nee-ah (Lightning) was lofted into a high ellipti- cal orbit on April 23, 1965; apogee 24,470 miles in the Northern Hemisphere, perigee 309 miles in the Southern Hemisphere. Its initial television relay was from the Pacific port of Vladivostok to Moscow about one-third the distance around the earth.

Molniya 2 took to the skies on October 14, 1965. News in December credited it with relay of the first color television from Moscow to Paris. By mid-1969 12 Molniyas had been orbited. Twenty-four receiving stations were reported to be lined up across the country to pick up the communications, whether radio, television, or telephone.

XI   Satellites Exploring for Science

> The more I see, the more impressed I am
> not with how much we know but how
> tremendous the areas are that are yet to
> be explored.
> —*John H. Glenn, Jr.*

ASIDE FROM COMMUNICATIONS SATELLITES, SPACE SCIENCE CALLS
for instrumented spacecraft designed to obtain scientific data on
space environment, the sun, earth, planets, and the galaxy—all essential to understanding the physical universe, which is vital to the
design of all space vehicles, including manned spacecraft.

The many objectives of these orbital laboratories and observatories
include meteorological research, weather forecasting, study of magnetic fields, radiations and gases, long-distance communications and
navigation, as well as gathering new knowledge of phenomena in the
atmosphere and outer space. They travel out across the solar
system, penetrating into regions that may long, or forever, deny
access to man. Equipped with instrumentation in various forms,
powered by solar cells and nuclear power, they amass scientific
information that sheds new light on "the new ocean." They too
communicate. By radio, telemetry, and television they reveal what
they hear and see.

PIONEERS

A series of probes designated as "lunar" and "deep space" were
instituted by NASA in cooperation with the USAF and the Army

beginning on August 17, 1958, when Pioneer 1 exploded after 77 seconds in flight. A second Pioneer 1 sent up on October 11, 1958, reached an altitude of 70,700 miles, and re-entered the next day.

The Pioneer scoreboard follows:

Pioneer 2, November 8, 1958, attained an altitude of 963 miles; the third stage failed.

Pioneer 3, December 6, 1958, climbed 63,580 miles and discovered a second radiation belt; re-entered December 7.

Pioneer 4, March 3, 1959, went into a solar orbit.

Pioneer 5, March 11, 1960, communicated data from 17.7 million miles; its radio signals were picked up from 22.5 million miles as it went into a solar orbit.

PROBES OF SOLAR SPACE

Pioneer 6, a 140-pound instrumented drum-shaped satellite, was projected into an elliptical path around the sun on December 16, 1965. On the eleventh day of its journey, when 1,241,039 miles from earth and moving 3,912 miles an hour relative to the earth's speed, radio from the spacecraft flashed evidence that the solar wind speed was 670,000 miles an hour, considered by scientists to be "relatively slow"! The signals indicated that "comparatively few charged particles and fairly unfluctuating magnetic fields" were encountered. Continuing on its flight, the spacecraft was expected to learn how far out the sun's "atmosphere" extends, and, therefore, where interstellar space begins. It passed behind the sun 167,591,700 miles from earth on November 19, 1968, and emerged four days later. Nearest approach to the sun was estimated at approximately 76 million miles.

Still intent upon learning the secrets of outer space, NASA released Pioneer 7 on August 17, 1966, to study solar weather. The drum-shaped 140-pound craft was scheduled to be about 12 million miles outside the earth's orbit in 28 weeks. Flying in a wide-ranging orbit between the earth and Mars it would take about 14 months to make a revolution of the sun. Through the measurement of particles emanating from the sun, such as cosmic rays and supersonic winds, scientists hoped to learn more about solar storms, radiation, and magnetic fields.

Their hopes were not in vain for Pioneer made multiple measurements. It found that the multi-million-mile-long magnetic tail of the earth—3.2 million miles away—may be blown off in huge chunks by the solar winds. And confirming evidence sent back by Pioneer 6, it detected that the wavering of the solar winds causes the gigantic magnetic tail to wag, or flap, like a flag in a strong breeze.

American 1967 launchings completed on December 13 featured a two-in-one shot from a Delta rocket that put 145-pound Pioneer 8 into a solar orbit on a mission similar to Pioneer 7. A 40-pound communications satellite "TTS" (Test and Training Satellite) was dropped off in earth orbit ranging from 200 to 370 miles.

Pioneer 9, a sun-orbiter, was launched on November 8, 1968, and at the same time a Delta rocket hurled into a 200-mile-high earthorbit a 40-pound communications satellite to serve as a test and training target for the global Apollo 18-station tracking network. Carrying instruments to record and transmit information about cosmic rays, solar winds, and magnetic fields, Pioneer 9 was aimed to swing around the sun once every 297.5 days at distances of 70 million to 93 million miles from the sun.

TIROS WEATHER WATCHERS

Tiros (Television Infra-Red Observation Satellite), shaped like a drum 42 inches in diameter and 19 inches high, took to the sky in April 1960. It was the first satellite to carry both radio and television, and is expected to remain in orbit for 50 to 100 years. The communication installation was operative for 78 days, during which time it sent back more than 22,000 pictures of cloud formations and panoramic scenes of the earth as televised from about 400 miles.

Built by the RCA Astro-Electronics Division, under the general systems management of NASA, Tiros comprised the most elaborate electronics installation ever sent into orbit up to that time. Equipped with approximately 450 transistors and semi-conductor devices, it also carried two miniature television cameras, video-tape recorders, infrared sensors, radio transmitters, 9,000 solar cells, and rechargeable battery-power supplies.

How Tiros Performed Speeding along its orbit in space, Tiros was linked to an extensive ground network of tracking and receiving stations, data processing systems, as well as programming and control centers. Together, the satellite and ground equipment formed a unified system to collect and analyze world-wide data on cloud formations in the earth's atmosphere.

At the start of each orbit the television cameras could be instructed electronically to photograph an area of specific interest, such as a typhoon center over the Pacific or a hurricane in the Atlantic. The instructions, prepared at the NASA Computing Center in Washington in cooperation with specialists of the United States Weather Bureau, were sent to the ground stations. At the appropriate station, the program was sent in the form of radio signals to an "electronic clock" in the satellite. The clock stored the instructions somewhat in the fashion of a remotely operated alarm clock, causing the cameras to start a sequence of operations at the specified time during the succeeding orbit, as the satellite passed over a region of particular interest.

Linked to each camera was a miniature television magnetic tape recorder specially designed for satellite use. Each recorder stored up to 32 individual pictures taken by the camera during an orbit. When the satellite passed within range of one of the ground stations, a command signal caused the information to be read from the tape into the satellite transmitter for transmission to the earth. For picture taking while the satellite was within range of the station, the cameras could be instructed to feed their information directly to the transmitter rather than to the tape storage system.

Tiros 2, a 280-pound television infrared weather eye, was hurled into space when a Thor Delta rocket thundered off the launching pad on November 23, 1960. Fourteen minutes after the take-off, 435 miles above the Atlantic, Tiros was freed from its protective fairing in the rocket's nose and was shot into orbit at more than 16,800 miles an hour by a powerful third-stage rocket. Within 100 minutes the drum-shaped satellite passed over the east coast of the United States to complete its first trip around the world, as ground stations picked up its televised pictures of cloud formations and measurements of heat levels in the upper atmosphere.

This weather-eye satellite was equipped so that ground observers could regulate its tilt by remote control and thus gain improved TV picture coverage. Another innovation was an infrared unit designed to measure portions of the infrared spectrum around the earth.

Two subminiature FM transmitters, about the size of a pocket radio, each weighing 33 ounces, were carried by both Tiros 1 and Tiros 2. These rugged radio transmitters, developed by Radiation, Incorporated, sent back the cloud pictures which were picked up by the 60-foot dish antenna system at Fort Monmouth, New Jersey. After seven months in orbit Tiros 2 continued to transmit pictures of cloud formations, in fact, more than 32,000.

A Hurricane Hunter Tiros 3, a 285-pound meteorological satellite, was launched by NASA in July 1961 as a hurricane hunter to detect storms spawning in the Caribbean and Atlantic. Basically it was the same as Tiros 1 and 2 except that one of the narrow-angle TV cameras was replaced by a wide-angle camera to operate in sequence for taking pictures over a 562,000-square-mile area as the satellite encircled the earth every 98 minutes in a 400-mile-high orbit.

Its main assignment was to send back pictorial clues relative to the birth of tropical disturbances and follow them from the embryonic state to the end of their violent whirl. To accomplish its mission, this Tiros, in addition to its TV eyes, carried a magnetic tape recorder to store up to 32 pictures during each orbit for transmission when within a 1,500-mile range of ground stations at Wallops Island, Virginia, and San Nicholas Island, off California. More than 9,000 solar cells were aboard to charge the chemical batteries and keep the electronic instruments operating for at least three months.

Tiros 4 went into orbit on February 8, 1962; Tiros 5, on June 19, 1962; 6, September 18, 1962; 7, June 19, 1963; and 8, featuring a new TV camera and automatic picture transmitting system, was launched on December 21, 1963. These satellites had a specified requirement of 90 days' operation but exceeded that period two or three times over. In providing a continuous weather watch since 1960 they transmitted more than 330,000 weather pictures.

A drum-shaped Tiros 9, orbited on January 22, 1965, was engineered to turn on its side and roll like a wheel in a polar orbit as two TV cameras scanned a continuous swath of the weather over the sun-illuminated areas of the earth. While it encircled the globe every 119 minutes—perigee 436 miles, apogee 1,602 miles—the TV pictures were stored on magnetic tape for reading out on command when the weather watcher was within range of a ground station. Incidentally the "cartwheel" was an experimental version of a Tiros Operational

System (TOS), six satellites planned to provide global meteorological data to the Weather Bureau on a daily basis.

Tiros 10 took to a polar route on July 2, 1965. It was Tiros 7, however, orbited in June 1963, that won the title "patriarch of the Tiros family." At the beginning of its third year in space it had traveled 275 million miles and had sent back 110,000 pictures during 10,500 around-the-world trips. It was turned off in May 1968 after 59 months of useful life. Tiros 10 operated for 24 months.

ESSA WEATHER WATCHERS

As the initial link in a global operational weather-satellite system, a 305-pound storm hunter called Essa 1 (Environmental Science Service Administration) was orbited on February 3, 1966. Traveling in a north-south path ranging from 433 to 523 miles high, it had two TV cameras pointed to scan the surface of the globe once a day. The pictures stored on magnetic tape to be radioed on command to ground stations comprised a report called a "nephanalysis" for use by the Weather Bureau. After 28 months of useful performance Essa 1 was shut off in June 1968.

Other Essas were planned to expand the nephanalysis and keep it pictorially up to date as the dual TV cameras focused on the weather as seen from on high. Essa 2, reported to have cost $2.6 million, was lofted on February 28, 1966, into a north-south path that varied from 843 to 885 miles high.

Essa 3 went up on October 3, 1966, to look over the earth's cloud cover from altitudes ranging from 859 to 923 miles. By the time Essa 4 was rocketed on January 26, 1967, receivers were available in 28 countries to detect the pictures. Essa 5 joined the array on April 20, 1967, to maintain the stream of weather pictures of which more than 800,000 had been recorded since Tiros 1 inaugurated the weather watch in 1960. Essa 6 took off on November 10, 1967, to eye the earth from a polar orbit between 874 and 922 miles high. Its photographs estimated to cover a 4.5 million square mile area were made available to weather stations in the United States and 45 foreign countries.

Essa 7, orbited on August 16, 1968, from Vandenberg Air Force Base, was stricken with a blinded TV "eye" relegating the 320-pound craft to an inoperational status.

Essa 8, weighing 290 pounds, joined the weather-watch line-up on December 15, 1968, with two TV cameras to look at the cloud cover every six minutes from a 900-mile-high circular orbit around the poles.

Essa 9 lifted off on February 26, 1969, aimed at a 900-mile-high orbit from which its TV cameras would survey weather conditions of the entire globe once every 24 hours. The 320-pound, drum-shaped craft supplemented three other such weather scouts regularly transmitting cloud pictures. It was the twenty-first weather satellite launched by the United States since 1960. By 1970, the Tiros-Essas had transmitted close to 1,300,000 pictures, and six of these weather-watchers were still in operation. They would be supplemented in 1970 by Tiros M, a 675-pound craft described as "equal to two Essas plus" since it would provide global coverage every 12 hours instead of 24 as with an Essa.

Nimbus   As a further step in development of a global meteorological satellite system, NASA placed a contract with the General Electric Company to build a satellite named Nimbus, described as a most sophisticated structure. The first Nimbus in a planned series of four, estimated to cost $241.7 million, was orbited on August 28, 1964. A trio of TV cameras soon were scanning hurricane Cleo along the Florida coast. At nighttime infrared sensors obtained pictures of the clouds and earth. Geographical views were amazingly clear and weather observations were described as "the most detailed pictures ever seen." Radiometer maps revealed that bands of warm air circulate inside a spiraling hurricane and cold air outside the storm. The mission ended in three weeks.

Nimbus 2, ranked as "the most advanced of the 14 weather satellites orbited so far by the United States," was launched on May 15, 1966. Weighing 912 pounds and equipped with four TV cameras, it circled the globe every 108 minutes in a south-to-north polar orbit ranging from 687 to 726 miles high. Engineered to aid in forecasting weather two weeks in advance, it could take about 3,000 weather pictures daily and transmit them to any of 150 automatic picture transmission (APT) stations. The earth's heat balance measured by infrared heat sensors informed meteorologists about storms in development. Thousands of solar cells mounted on two wing-like paddles supplied electric power.

During its first six months in space Nimbus 2 transmitted 860,000 weather pictures, including 16 typhoons and 9 hurricanes. It completed 2,449 orbits and traveled 72,000,600 miles on its mission "to scan every acre of the earth's surface" from an altitude of about 700 miles. In mid-1968 it was still transmitting and had sent more than one million pictures.

Nimbus 3, a 1,260-pound weather watcher, was equipped with two nuclear-powered generators and plutonium reported to have been worth $500,000. Within a few minutes after blast-off from Vandenberg Air Force Base on May 18, 1968, it went awry and was destroyed, falling in the Pacific about 95 miles off the California coast.

Another weather eye, given the name Nimbus 3, went into space on April 14, 1969, designed to specialize on in-depth measurements of temperature and humidity within the atmosphere from a 600-mile-high orbit. It was powered by solar cells and two 25-watt nuclear (SNAP-19) generators. Meteorologists who analyzed the radioed observations from the infrared sensors appraised the satellite as "a dramatic success."

VERSATILE ATS SATELLITES

A new combination weather watcher and radio-TV relay satellite was introduced by the United States on December 6, 1966. Known as a "Application Technology Satellite" (ATS-1), the 775-pound barrel-shaped craft went into a "stationary" orbit 22,300 miles above the Pacific, from which position it could relay communications among three continents, also "talk" to aircraft and televise weather over one-third of the world. Preliminary tests featured conversations relayed between a ground controller at the Goddard Space Flight Center, Greenbelt, Maryland, and airplanes scattered around the world. Revealing its versatility from the vantage point near the intersection of the Equator and the International Dateline, color TV was relayed from California to the East Coast, cloud patterns and storms were photographed over eastern Asia, the Pacific, and most of the United States.

A jet over the Pacific used ATS-1 in March 1967 to test the use of satellites in relaying aircraft navigation data. When a Tokyo-bound Pan American World Airways cargo plane was nearing Wake Island, the captain pressed a button and two green lights blinked on a console at Kennedy Airport in New York indicating the jet's position 6,500 miles away.

The pilot determined his position with ordinary navigation aids, set the numbers on a transmitting device, pressed a button to encode the data, and radioed it to New York via the space agency antenna in the Mojave Desert, California. From there the signals were routed over telephone lines to New York via the Flight Center at Greenbelt, Maryland. Aviation engineers also talked with the captain via the satellite hovering high above Christmas Island.

Success of the experiment presaged development of an instant, automatic, and continuous plotting of supersonic plane flights over the ocean area. "When supersonic planes are traveling 30 miles a minute," said an engineer, "you'll need satellite communications to be able to make quick decisions."

Described as "a giant daddy-longlegs spider" because of its 251-foot booms, ATS-2, a $9-million craft launched on April 5, 1967, was the victim of a misfiring rocket engine that was designed to put it in a 7,000-mile-high circular orbit. Plans had called for it to whirl around the globe every 6½ hours with two TV cameras scanning the weather as well as the earth.

The third ATS in a planned series of five to be built by the Hughes Aircraft Company was launched on an Atlas-Agena rocket on November 5, 1967. The $18-million spacecraft was "parked" in an orbit 22,300 miles above Brazil, thus providing a new link in a satellite communications system "that everybody can use"—a sort of global party line estimated to cost $178.1 million. As features ATS-3 was equipped to photograph the earth in color, "talk" to airplanes, aid in weather forecasting, and perform other functions related to navigation and communications.

Samos, Midas, and Others   Samos (Satellite And Missile Observation Station), the first in a series, was launched in January 1961 as an orbiting watchdog, which sees everywhere from its polar orbit, since it carried test equipment described as telescopic for reconnaissance photography from 350 miles above the earth.

Midas 1—a sentry in space—(Missile Detection Alarm System) failed to go into orbit in February 1960, but three months later Midas 2 was rocketed into a near-equatorial orbit. Midas 3, was hurled into a 1,850-mile-high polar orbit from Point Arguello, California, as an experimental forerunner of a satellite system that could detect the hot exhaust of a missile launched anywhere on earth. Two Midas satellites sent up in October 1963 were reported to be outstanding successes.

Lofti—an acronym for Low Frequency Trans-Ionospheric satellite—was launched by the U.S. Naval Research Laboratory early in 1961 to aid in the study of how very low frequency radio signals (3 to 30 kilocycles) interact in the ionosphere. Before it burned up in the atmosphere, after about six weeks, the data sent back confirmed belief that the ionosphere is not as opaque at very low frequencies as generally assumed, and that these waves do pass through the ionosphere with little attenuation and travel into outer space.

Oscar (Orbiting Satellite-Carrying Amateur Radio), a transmitter broadcasting "hi" in the radio code, rode piggyback on the Discoverer 36 satellite launched from Vandenberg Air Force Base, California, on December 12, 1961. Before the ten-pound Oscar completed its three weeks in orbit, 570 radio amateurs in 28 countries reported hearing the signal. A series of Oscars followed; Oscar 4 was dated December 1965.

Pegasus, a 3,200-pound satellite named for the winged horse in Greek mythology that ascended to live among the stars, was sent up on a similar mission on February 16, 1965. It featured a 96-foot wing spread that opened when it got into orbit; apogee 468 miles, perigee 308 miles. Within that range its task was to measure potential hazards of meteoroids to astronauts as well as spacecraft. Pegasus 2 went up with the same objective on May 25, 1965, and No. 3 winged into space on July 30, 1965.

ORBITING OBSERVATORIES

Geophysics and astronomy are the basis for a substantial number of orbiting observatories intended to aid in the study of phenomena in earth-space environment and in solar research. The physics program relates to radiation belts, magnetic fields, the ionosphere, solar winds, and interplanetary plasmas. Through chemistry the objective is to gain new knowledge pertaining to the dynamics and composition of the atmosphere.

Already from radio astronomy and rocket observations, scientists report having learned that the corona of the sun is not a steady, homogeneous structure but a fluctuating, loose collection of clouds of highly ionized gas. And rocket measurements of X-rays from the corona indicate that temperatures in the corona range from 600,000 to 2,000,000 degrees centigrade, much hotter than the surface of the sun. It is expected that future solar observatories and rockets will communicate new insights into interstellar processes, including the spectrum of radiation from the corona and neutron stars.[1]

Neutron stars, called "the most solid objects ever observed," have come under rocket observation. Such stars, made up entirely of closely packed neutrons, are believed to be about 10 miles in diameter, yet are estimated to weigh from 10 billion to 100 billion tons per cubic inch. Predictions are made that many millions of neutron stars may exist in the earth's galaxy, but are so hot as to be invisible; for example, billions of degrees Fahrenheit at the interior and millions of degrees on the surface, too hot to emit visible light. But they do radiate very short X-ray wavelengths, which have been intercepted by X-ray detectors rocketed high above the earth's atmosphere, which those rays do not penetrate.

GAZING AT THE STARS

A $69-million Orbiting Astronomical Observatory (OAO), the first in a 4-flight, $378-million series was tossed into a 500-mile-high orbit on April 7, 1966. Featuring a 21-foot wing spread and weighing 3,900 pounds, it was the heaviest, most complex electronic unmanned spacecraft launched in the United States up to that time. Equipped with a cluster of seven telescopes, its assignment was to stargaze uninhibited by the earth's atmosphere. Other devices on board were to measure wavelengths in the gamma-ray and X-ray regions of the spectrum. After two days in space, battery failure and other electrical malfunctions ended the mission which had been likened in historic importance to Galileo's first telescope.

The failure led to a redesigned version of a $75-million, 4,400-

---

[1] The neutron is an electrically neutral particle of unit mass composed of a proton and an electron closely united to each other.

pound, "most sophisticated, most automated, unmanned scientific satellite," launched on December 7, 1968. From a vantage altitude of 480 miles above the earth 11 telescopes focused on the stars, thousands of which were never observed with ground-based telescopes. Cylindrical in shape, OAO-2 measured 10 feet long and 7 feet wide and in orbit it had a 21-foot wing spread after the solar-power panels were deployed. While the telescopes were aimed at the stars, some of them from 450 to 2,000 light-years from earth, TV cameras sensitive only to ultraviolet light photographed them, and the electronic pictures were telemetered to earth.

OSO (Orbiting Solar Observatory), a complex 450-pound spacecraft, went into orbit on March 7, 1962. Its mission: to probe sunspots, X-rays, gamma rays, radiation, and other solar activities as it looked at the sun from an orbit averaging 355 miles high and unhindered by the earth's atmosphere. After 1,138 orbits the spin controls malfunctioned, and it spun so fast that the aiming devices could no longer focus the instruments and solar cells on the sun.

Automatic efforts to orient the satellite consumed so much power that the batteries were continually discharged and the radio signals ended. In its 77 days of operation about 1,000 hours of solar data were transmitted. OSO was equipped with 13 observing instruments, including a dust counter that watched the sun to determine whether microscopic dust came from its direction. A neutron counter sought to catch neutrons bounced up from the earth's atmosphere by the impact of cosmic rays. Eight such spacecraft were planned at a total cost of about $129.9 million, including the vehicle, launch, experiments, and data processing.

When OSO-2 lifted from the launch pad on February 3, 1965, the objective was to observe the sun's corona and obtain radiation data as it traveled at altitudes from 343 to 393 miles above the earth. Assigned to a similar "sun watch," a 727-pound OSO-3 equipped with nine telescopes and sensors went up on March 8, 1967, to scan the heavens from an orbit 354 to 377 miles high. Its gamma-ray telescope was reported to have detected high intensity of gamma rays streaming from the direction of the center of the Milky Way.

Listed by the Air Force Defense Command as the three-thousandth man-made object to be rocketed into space, OSO-4, designed to aid in development of a radiation warning system for astronauts, was put into a circular 350-miles-high orbit on October 18, 1967. It carried 250 pounds of instruments, including an X-ray telescope, X-ray spectrometer, and ultraviolet heliograph.

OSO-5, a $12-million, 641-pound satellite, was equipped with telescopes and sensors to measure radiation and to investigate solar flares. It was rocketed about 350 miles above the earth on January 22, 1969. OSO-6 followed on August 9, 1969, in a circular earth-orbit for study of solar flares and sunspots.

OGS (Orbiting Geodetic Satellite), a 350-pound, 36-inch spherical vehicle named "Anna," was launched on October 31, 1962, into a 700-mile high orbit; its mission: to help determine with outstanding accuracy the size and shape of the earth. In effect it was a man-made winking star by virtue of its four high-intensity lights that flashed periodically 20 times a day for observation by astronomical telescopes.

OGO (Orbital Geophysical Observatory), weighing 1,073 pounds, was space-borne on September 5, 1964. Its dragon-fly-shaped frame was packed with instruments to probe a wide scope of phenomena in the atmosphere, magnetosphere, and interplanetary regions. It went into a desired oval orbit extending 93,000 miles, but some of the devices malfunctioned and results were reported to be disappointing.

It was not until October 14, 1965, that OGO-2 was orbited with instruments to measure radiation. Its objective was thwarted, however, when for some reason or other it tumbled in space and the electrical power supply was practically depleted.

Next on the agenda was OGO-3, a 1,135-pound, bug-like observatory introduced from Cape Kennedy on June 6, 1966. Reputed to be carrying more scientific instruments than any previous United States spacecraft, it was scheduled to perform at least 21 experiments. It soared in a long, looping orbit 170 to 75,768 miles high, making one revolution of the globe in two days. It probed for information about various factors in the relationship of the earth to sun, i.e., solar winds, flares, magnetic fields, radiation, proton storms, aurora, ionization, and other forces that could jeopardize spacemen in flight.

The year 1967 was not without an OGO, for on July 28, OGO-4, a 1,240-pounder accredited as "a space laboratory," went into a 256-

by-564-mile-high polar orbit to take radiation measurements. At the time it was estimated that six flights by OGOs would cost about $266.1 million.

OGO-5 joined the cavalcade on March 4, 1968, equipped with instruments to conduct an extensive study of the relationship between the earth and sun. It moved in an egg-shaped orbit ranging from about 175 to 92,000 miles distant from the earth.

NAVY NAVIGATION SYSTEMS

The idea of navigation by satellite originated in 1958 when scientists at Johns Hopkins University observed that the position of the first Sputnik could be plotted precisely by using radio signals emitted by the satellite. Such "lighthouses" or "radio stars" appeared to be ideal for all-weather navigation by ships, submarines, and aircraft.

The U.S. Navy, which pioneered loran (Long Range Navigation) during World War II as a system for determining positions of ships and planes, took up the new idea to develop a space navigation system. Its first experimental navigational satellite, Transit 1-B equipped with four radio transmitters, was orbited on April 13, 1960. Eventually descending from its original altitudes ranging from 206 to 410 miles to 162 to 197 miles, the 265-pound, 3-foot sphere was reported to have dropped out of orbit in October 1967.

In comparison with the satellite system, loran measures the difference in time of the arrival of accurately synchronized radio pulses from special transmitting stations on shore. Loran also differs from radar in that no transmission from the ship is needed and no echo is involved. Furthermore, unlike radio direction finding, loran measures time of arrival rather than direction of arrival of radio signals. Skippers praising loran occasionally expressed doubt that it was a cure-all for ocean navigation; they accepted it as an aid to navigation to be utilized in conjunction with other available means. Nevertheless, during World War II about 3,000 ships and 30,000 aircraft used loran as "an excellent and reliable system."

Then came a new method—the Navy's satellite navigational system—undreamed of in the world war years. When operation began in 1964 the system was greatly enhanced by computers for precise and automatic calculations that threatened to outmode previous methods of sea and air navigation.

Four ground stations track the satellites as they pass within range: Wahiaha, Hawaii; Point Mugu, California; Rosemont, Minnesota; and Winter Harbor, Maine. From the tracking data, a computer center at Point Mugu makes quick, precise calculations of the satellite's position for the next 16 hours. The information is radioed to the satellite to be stored in its computer memory unit. The satellite's transmitter takes the data and beams coded messages every two minutes, thus revealing its location.

The frequency of the radio signals rises and falls depending on the nearness of the satellite to a ship. The effect is like the Doppler shift, which causes a waxing and waning in the whistle pitch of a passing train. Analysis of the shift, the pattern of which is fed into a computer, makes it possible to determine the moment when the satellite is nearest to the ship, and its distance at that time. The combination of the coded messages and the Doppler effect tell a ship's navigator within a fraction of a mile, through computer calculations, exactly where the ship is located.

The system was classified and restricted to the Navy until 1967 when it was announced that permission would be granted to commercial ships and ocean researchers to be equipped to tune in the coded signals. Ships can plot their course with an accuracy of about 600 feet, permitting them to reduce voyage time up to 10 per cent. Conventional navigation systems which are sensitive to weather have an accuracy of two to four miles. Engineers estimate that the satellite system especially increases utilization of oil tankers which spend 75 per cent of the time at sea and cargo ships generally more than 50 per cent. Shortened voyages resulting from accurate navigation are achieved, also significant reductions in man-hours required for shipboard navigation tasks.

Such a system is a boon to oceanographers who must know their exact position in mapping the ocean floor and in pinpointing discoveries. Pioneering full-fledged civilian use of space navigation for oceanographic research, the "Vema," a 202-foot, three-masted schooner of Columbia University, while probing the depths of the North Atlantic in 1966, took fixes of the Navy's Transit satellites as each passed overhead every 90 minutes. The ship's true position was determined and other precise measurements in the ocean survey were made. Heralding a new era in oceanography and marine geophysics, scientists reported that the navigational error with the satellite-computer, all-weather system was reduced to yards, whereas error in conventional navigation using the sun, stars, and land-based electronic signals may involve miles.

OMEGA

The Navy also developed Omega, an all-weather, general-purpose navigational aid for aircraft, ships, and submerged submarines. It continuously provides a position to a navigator in any weather; operation of the receiver is quick, simple, and reliable.

Four transmitting stations for the Omega system, located in Bratland, Norway; Port of Spain, Trinidad; Haiku, Hawaii; and Forestport, New York, provide navigational coverage to an area of more than one-fourth of the world. Four additional stations are planned for service in the areas of the Western Pacific, Tasmanian Sea, Indian Ocean, and southern South America. Early in the 1970s, therefore, it is expected that the system will provide continuous world-wide position information, accurate to one mile in daytime and two miles at night.

While loran is effective up to about 1,000 miles from shore, the Omega very low frequency signals that bounce between the ionosphere and the earth can be detected up to 8,000 miles under all weather conditions, day and night. Furthermore, such waves are not affected by normal variations of the ionosphere and other interference that affect the high frequencies.

The superliner "Queen Elizabeth 2" is equipped with all forms of modern navigational systems including loran, satellite navigation, and an Omega navigation receiver praised for accuracy and dependability in telling the QE2's position within a mile at all times.

Triplets Three satellites traveling piggyback atop a rocket, like a triple scoop ice cream cone, were aimed at separate orbits 550 to 629 miles above the earth on June 29, 1961, by the U.S. Navy. Designated as Transit 4-A, the principal unit weighing 175 pounds was heralded as the first operational prototype and experimental forerunner of a satellite navigation system. It carried the first atomic battery into space. Developed by the Atomic Energy Commission, it was named SNAP (Systems for Nuclear Auxiliary Power). 4-A on its sixth anniversary in space continued to signal data on command, and as the oldest operating satellite had traveled 868 million miles.

Greb, the second vehicle, was sent aloft to measure X-ray radiation from the sun and its effects in causing ionospheric disturbances and radio blackouts on earth.

Injun, the last of the trio, a 45-pound, drum-shaped satellite, was designed to facilitate study of intense radiation in the Van Allen belt, and its relation to the aurora borealis.

The triplets are expected to remain in orbit for at least 50 years, although Greb and Injun failed to spring apart and functioned at reduced efficiency. Transit, however, with its small nuclear battery, or generator, the output of which was only 2.7 watts, was reported to be working perfectly, supplying just enough power to operate two of the four radio transmitters, thus supplementing the solar batteries.

Five months later Transit 4-B was sent into space with the same type of nuclear unit—about the size of a grapefruit. Something went wrong with it in mid-1962, but the nuclear battery aboard Transit 4-A was reported to be still generating electricity in mid-1963. Transit 5-A, a navigational satellite launched in December 1962, was reported "electronically dead."

Equipped with a 27-pound nuclear generator as its sole electric power source, Transit 5-B was orbited on September 28, 1963, from Vandenberg Air Force Base. Named SNAP 9-A, the generator was rated as a major advance toward eventual elimination of conventional chemical batteries and solar cells on satellite and rocket projects.

The radioisotope-fueled generator, with no moving parts, was designed to produce a steady flow of 25 watts, sufficient to power the satellite's radio transmitters, data recording apparatus, and other electronic instruments. The energy is derived from normal radioactive decay of a lump of plutonium 238 that creates heat which is converted into electricity by a thermocouple comprising two strips of dissimilar metals. Estimates put the life span of such nuclear generators at about 250 years.

The first nuclear reactor, SNAP-10A, designed to operate in space, was launched on April 3, 1965. Circling the globe on a path over the North and South poles every 112 minutes, it traveled at an altitude of

800 miles. As a 250-pound miniature power plant built to produce a steady flow of 600 watts, scientists calculated it might remain in orbit for 3,500 years.

Also on board for test was a 2.2-pound ion propulsion engine designed to manufacture its own power by electrically vaporizing cesium into atomic particles called ions and expelling them at high speed through a nozzle to provide thrust. Although the thrust was only two-thousandths of a pound, it was rated as sufficient in interplanetary vacuum to push a spacecraft up to speeds of 100,000 miles an hour. After encircling the earth for 43 days the reactor failed to function.

Secor (Sequential Collation of Range), a navigational type satellite was instituted by the U.S. Army in 1964, the main objective being to remap the Pacific islands. Within the next two years six such spacecraft were sent up to bounce back high-frequency radio signals transmitted from fixed ground stations. The angle of the radio beams and the time required for the signals to complete a round trip made it possible to measure the accurate position of Secor. Then surface triangles could be plotted precisely. For example, it was determined that Japan's Ryukyu Islands were half a mile southwest of where maps had located them.

A mapmakers' satellite Pageos 1 was launched on June 23, 1966, from Vandenberg Air Force Base. The 125-pound, aluminum-coated plastic balloon went into a circular polar orbit 2,649 miles at the highest point and 2,616 at the lowest, circling the earth every three hours at 13,700 miles an hour. At night it appeared as bright as the North Star, as sunlight reflected from the shiny surface. More than 40 widely separated ground stations could photograph it for plotting purposes. The resulting pattern of triangles enabled mapmakers to spot any point on earth within 100 feet of the true position.

Eros (Earth Resources Observation Satellite), proposed by the Interior Department, would carry three TV cameras to map the United States in less than three weeks, whereas aircraft would need an estimated ten years to complete such a photographic mission. The earth's natural resources—water, oil, mineral, geographic, agriculture, etc.—would be featured in the scanning, which scientists hoped to begin during the 1970s.

Explorers   America's first satellite Explorer 1 put the United States in the space race in January 1958. It was acclaimed as "a great satisfaction to everyone who worked on it" as it continued to orbit the earth in 1966.

From 1958 on there was no letup in Explorer launchings; some failed to orbit, several re-entered the atmosphere, others sent back scientific data, while a number continued in flight through space with life expectancy ranging from 5 to 20 years.

Explorer 12, an 83-pound, paddle-wheel satellite, was launched by NASA in August 1961 for a study of solar "energetic particles" outside the influence of the earth's magnetic field. Within 13 hours it had traveled 48,500 miles into space, probing with a 32-inch boom at the end of which was a magnetometer for measuring the strength of magnetic fields. As the satellite swept down within 180 miles of the earth and then started out again on its 27-hour round trip, radio flashed the data so that scientists might gain basic knowledge on what man must encounter in the radiation belt he must navigate to reach the moon and planets.

To study the lethal effect that space dust could have on astronauts, a Scout rocket hurled Explorer 13 into an orbit ranging from 606 to 175 miles from earth. The mission was to detect micrometeoroids, dust-like particles believed to result from an exploded planet; such remnants, traveling 45 miles a second, could erode or puncture a spacecraft. The 127-pound Explorer, shaped like a cylindrical water cooler, sped around the earth at 17,500 miles an hour, once every 92.27 minutes. Launched by NASA on August 25, 1961, it was the fiftieth satellite placed in orbit by the United States since the first Explorer 1 was launched in January 1958. Scientists have expressed delight with the data from the Explorers, which have contributed new knowledge of the atmosphere.

Explorer 15, a 98-pound satellite, went aloft on October 27, 1962, to survey artificial radiation created by high-altitude nuclear explosions over the Pacific. It traveled 20,000 miles an hour in an orbit ranging in altitude from 170 to 10,360 miles.

Explorer 16, orbited on December 16, 1962, was described as a micrometeoroid satellite. During its first month in space there were indications that 11 micrometeoroids punctured the trim beryllium-copper pressure cells that comprised its skin. This was said to be the first time that actual punctures had been recorded; previously only non-penetrating impacts had been measured.

During a 17-month period, 55 meteoritic punctures registered in patches of its skin made thin to assess such hits. Within the satellite behind each patch was a tiny compartment of compressed gas. When the skin was punctured release of the gas pressure was reported by radio. Analysis revealed slight danger of puncture from meteorites, according to Dr. Fred L. Whipple, head of the Smithsonian Astrophysical Observatory, Cambridge, Massachusetts. The orbit was between 470 and 740 miles above the earth.

Explorer 17, a 410-pound instrumentated satellite, was vaulted into an egg-shaped orbit from 150 to 570 miles high on March 2, 1963. Traveling around the globe every 96 minutes, its eight measuring instruments encased in the satellite's stainless steel shell recorded the amount of helium, oxygen, and nitrogen encountered as well as other information relative to the earth's atmosphere.

Discovery that the earth is surrounded by a belt of neutral helium at an altitude from 150 to 600 miles was said by scientists to add a new and unexpected complexity to the composition of the upper atmosphere. It had long been believed that the earth's atmosphere was a band of oxygen and nitrogen that gradually blended into the hydrogen atmosphere of interstellar space, but based on observations during the International Geophysical Year (1957–1958), scientists concluded that a broad helium layer exists between the oxygen and hydrogen. The probes by Explorer 17 were said to support their theory.

Geodetic Explorer 26, a 101-pound, windmill-shaped scientific satellite, whirled into orbit on December 21, 1964, to investigate radiation, including the Van Allen belt. It sped around the globe every 7 hours and 40 minutes in an elliptical path ranging from 190 to 16,280 miles above the earth.

Geos, a 350-pound geodetic Explorer listed as No. 29, soared upward from Cape Kennedy on November 6, 1965, to take the earth's measurements in an experiment aimed to produce the world's most accurate maps, pinpoint long-range missile targets better, and establish guideposts for tracking astronauts. Its altitude ranged from 700 to 1,300 miles.

Four high-power flashing lights could be triggered by ground command. The flashes photographed by telescopic cameras against a background of stars from a number of angles and simultaneously from different continents—plus precise radio, navigation, and tracking devices—were designed to help geodesists calculate distances on the globe and the general shape of the earth's surface as well as the strength of its gravitational field. Also attached to Geos were 322 prisms to reflect the beam of a laser mounted on an Intercept Ground Optical Recorder (IGOR) telescope at NASA's Goddard Space Flight Center, Greenbelt, Maryland, to explore the utilization of lasers in optical radar and space tracking. By measuring the time required for the laser light beam to make the round trip the spacecraft's exact position could be determined. Similarly equipped for laser experiments were geodetic Explorers 22 and 27 launched on October 10, 1964, and April 29, 1965.

Lunar Explorer (designated No. 35) went into space on July 19, 1967, scheduled for a 2- to 3-year mission to probe around the moon. New and significant findings reported by NASA indicated that the moon is "a cold, non-magnetic, non-conducting sphere" as analyzed from an orbit between 200 and 4,800 miles high. Also it was noted that "because of a lack of magnetic field, no radiation belts surround the moon, nor is there any evidence of the existence of a lunar ionosphere."

Continuing the mapping expedition Geos 2, launched on January 11, 1968, traveled around the globe once every 112 minutes in a path ranging from 920 to 630 miles above the earth.

Then came Radio Astronomy Explorer-A (RAE-A), a spider-like craft designated as Explorer 38, rated as a masterpiece of scientific ingenuity. It was the first of two satellites reported to cost about $24 million, designed to provide astronomers with "a new window on the universe." Orbited on July 4, 1968, from the Western Test Range in California, this unique spacecraft was planned for study of low-frequency electromagnetic waves beyond the earth's atmosphere, which frequencies below ten megacycles rarely penetrate. With four antenna legs unreeled to form a giant X measuring 1,500 feet from tip to tip, the 417-pound craft would listen in as it traveled from 400 to 3,700 miles out from the earth. Scientists hoped that in foraging for information it would detect data on "radio noises" generated in the Milky Way, also monitor sunspots and solar eruptions as well as mysterious rays or bursts of "noise" from Jupiter and from storms swirling in the cosmos.

After stabilization in an almost perfect circular orbit—highest and lowest points 3,641 and 3,636 miles—the perforated metal tube antennas were deployed over a four-month period until the breadth of 1,500 feet was achieved by four ground-controlled maneuvers. The antennas were said to be as delicate as those of a moth, and for that reason a circular orbit was essential. Scientists explained that in such an orbit the gravitational field remains uniform, avoiding any twisting distortion of the antennas. Furthermore, the earth's gravity influence on the antenna system would keep the base of the spacecraft always pointed toward the earth. Small TV cameras monitored the position of the antenna tips so that controllers on the ground could "keep an eye" on them.

IMP (Interplanetary Monitoring Probe), designated as Explorer 18, was orbited on November 26, 1963, on a dual mission: to explore regions beyond the earth's magnetic field and to help in planning for protection of astronauts against perils of radiation between the earth and moon. Traveling in a far-ranging elongated orbit extending from 120 miles above the earth to about 122,800 miles, the satellite is expected to orbit for hundreds of years, although the electronic apparatus was designed to function for about a year.

Listed as "a 128-pound physics lab," another IMP was sent up on May 29, 1965. Traveling from 120 to 130,000 miles above the earth its main duty was to measure radiation.

INTERNATIONAL SATELLITES

Ariel 1, the first international satellite, was orbited on April 26, 1962, from Florida with NASA in cooperation with the United Kingdom. Featuring a 132-pound British packet of scientific instruments and powered by four solar-cell paddles, it was designed to gather information about the ionosphere, solar radiation, and cosmic rays. Among its first discoveries—big bulges of dense ionospheric particles jetting out at least 3,000 miles.

On the same day that Ariel went up, the United States in association with Japan lofted a rocket from Wallops Island, Virginia, to investigate the ionosphere. A small gold-coated sphere attached to a two-foot boom flipped out from the side of the rocket to aid in the probe of temperature, etc.

Ariel 2, the second cooperative British-American space effort, was launched from Wallops Island, on March 27, 1964, to measure distribution of ozone in the upper atmosphere. Ariel 3 was planned to measure the size and quantity of micrometeoroids in space.

Alouette, the first Canadian satellite, was orbited on September 28, 1962, with NASA in cooperation with the Dominion. Shot up to an altitude of about 600 miles from Point Arguello, California, its function was to take soundings of the ionosphere as it sped around the earth every 105 minutes, passing ten degrees from the North and South Poles, and transmitting on 1.6 to 11.5 megacycles. In 1969 it was referred to as "the oldest active satellite" as it continued to transmit data on solar activity.

Alouette 2, launched in California on November 28, 1965, traveled about 135 million miles in 4,037 orbits during its first year. The 32-pound craft responded to more than 11,200 commands and all experiments were reported to have been conducted as planned. Canada's third satellite, ISIS-1, orbited on January 29, 1969, from the Air Force western test range, also was aimed to explore the upper atmosphere.

The Dominion planned to establish a Telsat communications system to be operating in the 1970s to widen television coverage for the benefit of Canadians in remote regions, many of whom were out of TV range, or saw only a few hours a week of video-taped programs. About 30 ground stations would relay the telecasts far and wide across the northland reaching Canadians on the frontiers as well as Eskimo settlements.

Australia and India reported similar plans for use of communications satellites by the mid-1970s to bring people in remote areas into closer contact with domestic and world affairs, education, sports, and entertainment.

Britain, Canada, and Italy in launching satellites used American-built boosters. When France entered the "space club" on November 26, 1965, its own Diamond rocket was used for the boost from a site in Algeria's Sahara. After two days the 88-pound, radio-equipped capsule designated A-1 was lost in silence.

It was not long, however, before France again went into space. On December 6, 1965, Diapason, a 135-pound instrumented satellite designed for navigational study, was blasted upward by an American rocket from Vandenberg Air Force Base. A near-polar orbit was achieved at an altitude of about 490 miles.

France's third and more advanced satellite, a 50-pound craft named Diadem 1, equipped for laser light-beam experiments, was launched by a Diamond rocket from the Sahara on February 8, 1967. The low point of the orbit was 363 miles above the earth and the high about 573 miles. Diadem 2 went skyward on February 15 from the Hammaguir base in the Sahara in an attempt to chart the topography of the Mediterranean area.

The first Italian satellite in 1964 was followed by a second in May 1967 when a 285-pound San Marco was hurled into the African sky by a Scout rocket provided by the United States. It was the initial satellite to be launched from a seaborne pad—a platform off the coast of Kenya. The objective was "to measure the density of the upper atmosphere over the equator."

A 238-pound satellite called Heos, built by the cooperative effort of a ten-nation European Space Research Organization, was boosted into space by a three-stage Delta rocket from Cape Kennedy on December 5, 1968. The $12-million drum-shaped craft streaked into an orbit extending as far as 138,000 miles from earth. Scientific instruments on board were to monitor solar radiation and magnetic forces. It was the first American civilian space agency satellite rocket sold to foreigners who were reported to have paid NASA $3.75 million for the rocket and launching.

SOVIET SATELLITES

Orbiting of a long series of Russian satellites labeled Cosmos—an all-purpose name—was instituted on March 16, 1962. Cosmos 51 launched in December 1964 is believed to have been the first satellite equipped with automatic instruments to measure stellar ultra-violet radiation. The hundredth Cosmos went into space on December 17, 1965. By the end of 1969 the count was 317. While most of them were described as "instrumented," apparently they were planned to accomplish various missions from scientific exploration to recon-naissance.

Cosmos 110, lofted on February 23, 1966, served as a "scientific dog house" for 22 days. Two dogs, Veterok (Breezy) and Ugolyok (Blackie), were on board for 330 orbits, the apogee of which reached 560 miles within the lower section of the Van Allen radiation belt. Nevertheless, the dogs were reported to have come through the ordeal in fairly good shape, except that they suffered muscular reduction, dehydration, and loss of calcium in the bones, indicated by a high calcium count in the blood as well as increased amounts of phos-phates. Anti-infection properties were decreased. Confusion in re-adjustment to walking was evident. Dr. N. N. Gurovski, Soviet scientist, also told an international conference on space health prob-lems at Geneva that some damage to the liver and intestinal tract was noted. Both dogs, he said, were restored to good health and had normal litters after their flight. Throughout the journey a TV camera in the capsule enabled ground observers to keep an eye on the ex-plorers referred to as "Cosmodogs" and "Dognauts."

Two unmanned Cosmos, No. 186 specified as large enough to accommodate a five-man crew, launched on October 27, 1967, and No. 188, fired into space three days later, docked in orbit. When the link-up was established they flew in tandem for 3½ hours at altitudes ranging between 127 and 171 miles, circling the earth once every 88 or 97 minutes. Then, in response to the engine systems, they resumed separate paths. Both craft soft-landed several days apart in a preset area in Russia. In the meantime Cosmos 189 was launched.

The successful rendezvous and automatic docking were attributed to "space radio-technical means and on-board computers." Radio and television transmitted pictures of the linked vehicles as well as telemetric information to a network of ground stations. Russian TV presented a film of the linked spacecraft, their separation, and Cosmos 188 traveling off on its lone journey as viewed from a camera aboard No. 186.

As a prelude to "creation of large space stations" that would conduct multifaceted explorations of outer space, the cosmonautic accomplishment was heralded as an outstanding technical triumph. It established a new *first* in communications related to accurate navigation, electronic control, radar guidance, and complex maneuverability that proved to be the forerunner of link-ups to transfer astronauts from one space station to another.

A repeat performance interpreted as another advance toward manned space flight to the moon was staged by Cosmos 212 and 213 on April 15, 1968. The target spaceship Cosmos 212 was orbiting about 140 miles above the earth at a speed of 17,500 miles an hour when the two craft found each other by radar, maneuvered together, and docked automatically under computer control. During the rendezvous over the Pacific they flew together for 3 hours, 50 minutes, until separated by radio command from earth, a feat witnessed by those in tune with the Soviet TV network. After several days of scientific experiments the spacecraft were reported returned to earth.

Next in the line-up came Cosmos 214 and 215, followed by 216 on April 20. Listed as an automatic observatory, Cosmos 215 carried eight telescopes to explore mysteries of the universe free of interference of the earth's atmosphere. High point of the orbit was 264 miles above the earth and the low point 161 miles. After several weeks of exploration the observatory was brought back to earth credited with having collected and radioed a vast amount of information relative to rays, radiation, and magnetic effects.

Continuing to probe the ionosphere, Russia orbited an unmanned Intercosmos 1 in October 1969, and Intercosmos 2 in December, described as "a joint research project with six Eastern European nations and Cuba."

PROTONS AND ZONDS

On the day that Cosmos 71 through 75 were launched by a single rocket, it was announced on July 16, 1965, that the Soviets had fired a powerful booster rocket with an estimated 3-million-pound thrust that put an unmanned spacecraft Proton 1 into orbit as the heaviest payload—13 tons—ever launched up to that time.

Proton 2, lofted on November 2, 1965, was reported to be equipped with a 10-foot bank of instruments to measure energy of the spectrum and determine chemical composition of cosmic rays.

Proton 3, listed as "a heavy scientific space station," went into action on July 6, 1966, traveling in an orbit 118 to 391 miles above the earth. Its instruments were to measure cosmic rays, to seek such rays with energies up to 100 trillion electron volts, to probe radiation belts, and to search for quarks—subnuclear particles that exist theoretically.

Classified as "the largest automatic space station ever launched," Proton 4 went into space on November 16, 1968, loaded with scientific instruments reported to weigh 27,458 pounds. That made the total weight of the craft about 37,478 pounds. It circled the globe every 91.75 minutes at altitudes ranging from 158 to 318 miles.

Another Russian series inaugurated on January 30, 1964, featured two instrumented satellites named Electron, launched from a single rocket. Zond-1, another Soviet satellite, was orbited on April 2, 1964, "to further development of a space system for distant interplanetary flights."

Mars was the announced goal of Zond 2, launched on November 30, 1964. It was described as being equipped with a new-type rocket engine, or plasma accelerator. Failure in the solar cells that power the sensing instruments was blamed for lack of communications during most of the journey. No results were disclosed, although the craft was believed to have passed within 1,000 miles of the planet.

Two days after launch on July 18, 1965, Zond 3 photographed the far side of the moon as seen from about 7,000 miles. The pictures as recorded were transmitted to earth on July 29 when the spacecraft was reported to be 1.4 million miles away.

Zond 4, listed as an automatic research station, went into space on March 2, 1968. It was estimated to have reached an extremely high apogee before returning to earth a week later.

Zond 5, launched on September 15, 1968, made history on a 500,000-mile journey that featured a circumlunar flight. After arching around the moon within about 1,200 miles, the spacecraft was guided back to earth with uncanny precision.

Scientists pointed out that the craft followed a "free-return trajectory," which required aiming the vehicle to fly close enough to the moon to enable use of lunar gravity to draw it around and then thrust it back toward earth with increased velocity without depending upon any on-board rocket. Engineers explained that if the craft had come within 100 miles or so of the lunar surface, the gravity would have pulled it into orbit and it might have gone astray in interplanetary space instead of heading homeward.

After successfully navigating re-entry into the earth's atmosphere, a parachute ejected for a splashdown on the Indian Ocean about 2,500 miles east of Africa's southern tip, where a Soviet ship stood by for recovery. It was the first time that a spacecraft had been returned to earth from so far out in space. The accomplishment acclaimed as a masterful achievement and a fantastic technical feat was widely heralded as an historic breakthrough, pointing the way to eventual round trip manned flights beyond the earth's atmosphere. It was disclosed that the moonship's instruments brought back "a vast amount of scientific information," not only by means of radio but as directly imprinted in recording devices, therefore free of background noise that might contaminate radioed data. Turtles, flies, worms, plants, and seeds on board were said to show few significant effects.

Seeking additional information on near-lunar space, Zond 6 went unmanned on a trajectory toward the moon on November 11, 1968. Its mission was announced "to conduct explorations along the flight route and to test systems and instruments aboard." Photographs of the lunar surface were taken at distances of from 2,170 to 6,200 miles. It looped the moon carrying biological and botanical objects as did Zond 5 to investigate effects of radiation. Again reports indicated no significant changes were detected after a week-long flight and landing in a preset zone in Russia. Zond 7, unmanned, launched on August 8, 1969, orbited the moon for photographs and returned to earth.

BIOSATELLITES AND ASTROBUGS

America's first biosatellite (Bios), a 944-pound beehive-shaped spacecraft headlined as a modern Noah's ark carrying beetles, fruit flies, wasps, spores, frog eggs, plants, and other earthly specimens, was orbited on December 14, 1966. It was the year's finale of blast-offs from Cape Kennedy's launch pad.

Traveling around the globe in 47 orbits 195 miles high, the objective was to afford scientists opportunity to study how insects, one-celled animals, and wild flowers react to the environment of space; how radiation and weightlessness affect them. On the fourth day, the biological laboratory was scheduled to parachute to earth to be retrieved on the Pacific. Failure of the re-entry retrorockets, however, caused it to be "lost in space." Tracking stations detected the errant capsule traveling in an uncontrolled orbit from which it was destined to drift down into the atmosphere and burn up.

Biosatellite 2, carrying a payload similar to that of No. 1, was aimed into a 196-mile-high altitude on September 7, 1967. After a flight of 45 hours on 30 orbits, the "ark" was caught on re-entry at 12,000 feet by an Air Force plane 1,000 miles southwest of Hawaii. The experiment to check on biological hazards was reported to have yielded 71 per cent of the information scientists had hoped to gain. They noticed that bacteria multiplied more rapidly in the weightlessness of space than on earth, where gravity is an influence; also that plants require gravity to grow straight, and that some organisms suffer more radiation damage in a weightless condition than on earth.

AN ASTRO-MONKEY

Booked as "the most intensive study ever made of a complex living organism in space," a 14-pound monkey named Bonny took off on an earth-orbiting mission on June 28, 1969. Strapped in $55-million Biosatellite 3, the monkey was elaborately instrumented to be a communicator of biological data during 469 scheduled loops around the earth at an altitude from 225 to 245 miles. An array of sensors were installed to collect the data to be radioed to ground stations.

On the ninth day, during the 130th orbit, ground controllers detected "serious deterioration" in the monkey's condition and decided on an emergency splashdown. A copter picked up the 1,536-pound capsule 25 miles north of Hawaii and flew to Hickham Air Force Base, Honolulu. Twelve hours later the death of the monkey ended the project estimated at $92 million.

Plans for Slomar (Space Logistics Supply, Maintenance, and Rescue) are reported to include aerospace planes with long metallic fingers that can be manipulated by remote controls for repairing and servicing space vehicles in orbit.

Thousands of other satellites and spacecraft will follow the trails blazed by the first 500; increased in size and weight they will carry astronauts as well as all sorts of equipment. Others will roam far across the universe, robots communicating by radio and television and controlled by electronics from earth. The "Santa Maria," "Pinta," and "Niña" of the Space Age already are on the cosmic sea in search of new worlds and new knowledge. In the years ahead they will seem small and simple compared to the odd-shaped spaceships that will make news in the heavens between 1970 and the year 2000.

XII   Robot Communicators

> Science has taught the world skepticism
> and has made it legitimate to put every-
> thing to the test of proof.
> —*Oliver Wendell Holmes*

THE CONQUEST OF SPACE MOVES FORWARD ON TWO FRONTS:
manned spacecraft and unmanned instrumented capsules that make
historic investigations in measuring, probing, photographing, and
communicating. The robots featuring unmanned ingenuity under
radio control from earth can go out into space for millions of miles
oblivious to the hazards that confront humans. Robots can make the
long journey of months' duration to Venus and Mars that man hopes
someday to achieve. But the distance and the dangers that lurk along
the way may prevent him from ever seeing first-hand what the robots
detect, photograph, and televise.

The impressive feats of the instrumented American Rangers,
Mariners, Surveyors, Lunar Orbiters, and the Russian Luna vehicles
led to the question: why risk the lives of astronauts when scientific
probes can accomplish such remarkable results in exploration? And
as the instrumented artifacts become more human in computerized
performance, the unmanned will be even more adept in penetrating
celestial distances and environments beyond the limits of man's
endurance.

Nevertheless, the idea persists among scientists that man must see
for himself; he must personally experiment, test, and explore. Therefore, long into the future dramatic flights into the cosmos are likely to

continue as manned and unmanned supplement. At least until man becomes convinced, if ever, that he is not destined to land in faraway places on which robot Columbuses can soft-land and communicate their discoveries back to earth.

Ranger Satellites Ranger 1, a 675-pound moon-test satellite carrying an intricate array of electronic and mechanical devices, was launched on August 23, 1961, as an experimental forerunner of an instrumented spacecraft planned for launching and landing on the moon—the prelude to a manned lunar expedition. Developed by the Jet Propulsion Laboratory at the California Institute of Technology, the primary purpose of the Ranger mission in flight was to test basic components to be used in spacecraft destined for landing on the moon and later on planets. Also under test was the "parking orbit" technique, featuring a second-stage rocket as a mobile launching platform in space to get the payload into correct position before shooting it toward the moon.

The plan was to hurl the satellite into an extremely elongated orbit to a high point of 685,000 miles, in about 58 days to swing back to 37,500 miles, then to continue on an eccentric orbit in which its movements would be controlled by a novel electronic brain. The flight, however, fell short of expectations and was only partly success-ful, since the rocket booster put the satellite into a close-in orbit with an apogee estimated at 312 miles and a perigee of 105 miles. A similar fate befell Ranger 2, rocketed on November 18, 1961.

Ranger 3 a 727-pound craft estimated to cost $7 million was launched on January 26, 1962 as "a 20th-century masterpiece of ingenuity." It was designed to conduct a number of scientific experiments and to televise the moon before crash-landing on the lunar surface. Excessive velocity of the second-stage booster caused it to miss the moon by 22,862 miles, and it went into a solar orbit.

Ranger 4, equipped with radio, TV camera, seismometer, and other scientific instruments, was hurled toward the moon on April 23, 1962. Malfunctioning of the master clock and electronic system was blamed for preventing it from sending TV pictures and other data. After a flight of 64 hours the beep of its tracking radio ended as it reached the far side of the moon, where it was believed to have crashed.

Ranger 5, a 755-pound, $15-million spacecraft carrying radio, a

TV camera, and various scientific devices, was lofted on October 18, 1962, bound on a lunar mission similar to that of Ranger 4. After 8 hours and 44 minutes of flight the battery apparently ran down when it failed to get solar power to maintain its charge, and the mission ended.

The successive failures halted the Rangers for engineering re-examination. On January 30, 1964, Ranger 6 headed for the moon. When the 8-foot, 804-pound spacecraft wandered slightly off course on its 240,000-mile, 66-hour journey, radio impulses from earth maneuvered it back. Six TV cameras were on board to transmit pictures of the lunar surface in the ten minutes prior to impact. The crash landing was on schedule in the Sea of Tranquility but the camera system failed to function.

Ranger 7, with modifications in the command and TV circuitry to avoid the ailments that plagued 6, took to space on July 28, 1964. It traveled 243,665 miles in 68 hours, 35 minutes, and in the 19 minutes prior to impact at Mare Cognitum its six TV cameras transmitted 4,308 pictures of the lunar surface, which were received at the Jet Propulsion Laboratory.

Ranger 8, weighing 809 pounds and equipped with six TV cameras, was launched on February 17, 1965, to scan the moon in the area of the Sea of Tranquility. It traveled 234,000 miles in 64.9 hours and in 23 minutes prior to a crash landing sent back 7,137 pictures.

Ranger 9, rocketed on March 21, 1965, was assigned to photograph the crater Alphonous, which it did with astounding accuracy and clarity; 5,814 pictures were received. For the first time millions of television viewers saw some 200 of the pictures live as the Ranger's TV cameras functioned to perfection. The scenes were transmitted from the spacecraft to the big dish antennas at the Goldstone tracking station in California's Mojave Desert, which fed the pictures to the TV networks.

A commentator's report on the approach of the Ranger to the moon and his explanations of what was in view made it a dramatic program packed with excitement and realism as if it were an event being telecast on earth. As the spacecraft raced toward the lunar surface the commentator counted out the miles yet to go and the minutes remaining before impact. Exactly on the second planned, he exclaimed "Impact!"; the mission was completed and the TV screens across the nation went blank as far as Ranger was concerned. Thus concluded the nine planned Ranger expeditions to the moon, estimated to have netted more than 18,000 close-up pictures of the moon at a cost of $267.4 million.

MARINER'S RENDEZVOUS WITH VENUS

Mariner 2, a 447-pound satellite laden with scientific instruments and 9,800 solar cells, was launched on August 28, 1962, headed from Florida on a 190-day voyage toward the planet Venus.

The take-off was perfect, but ten hours later a computer calculated that the spacecraft must alter its course or miss Venus by about 233,000 miles. Scientists at the California Institute of Technology's Jet Propulsion Laboratory were in command, confident that by a radio impulse they could "nudge" the satellite back on course. So while it was traveling more than 60,000 miles an hour they flashed a signal across 1,200,000 miles to roll Mariner on its axis and snap its dish antenna out of the way of the mid-course motor exhaust. That cleared the way for another command to Mariner's electronic brain to execute a pitch maneuver, putting it in proper position to fire the craft's liquid-fuel rocket provided for just such a contingency—to shift the trajectory.

For some reason, however, the corrections rocket nudged the spacecraft into more velocity than planned. It only needed an additional 45 miles an hour to the 60,117 miles per hour it was traveling in relation to the sun. But the rocket lifted the added velocity to 47 miles and that could make quite a difference in its course. The scientists who were watching as it reached 7,330,000 miles from the earth and 30,200,000 miles from Venus pointed out that at interplanetary distances an overcorrection of just two miles was enough to alter the course of the craft by 10,000 miles. They figured, therefore that it would probably miss Venus by about 21,000 miles.

It was an outstanding illustration of man's ability to reach far into space by radio. All of the $10-million spacecraft's complex communications and navigation equipment was reported to be working perfectly as Mariner continued on its inquisitive mission.

True to the timetable, the satellite streaked by Venus within 21,648 miles on December 14. Tracking was so accurate that observers on earth knew within ten miles when Mariner reached that point. Traveling 85,000 miles an hour its Venus encounter lasted about 42 minutes. Then it went into an eternal orbit around the sun, presumably to continue until the end of time as a strange spider-like contraption out among the stars. On the day it sped past Venus, that planet was "the morning star," 35,900,000 miles from the earth. Astronomers said that if anyone had been aboard, Venus would have appeared as a brilliant orb 900 times the area of the full moon as viewed from the earth.

The radio signals from that remote region of the universe established a new long-distance record in communications. Equally fantastic was the performance of radio in finding the infinitesimal speck in order to track, maneuver, command, and flick instruments into action.

A Spendid Achievement Two days before Christmas, Mariner made its nearest approach to the sun, 65,539,000 miles; the gap between it and Venus was 2,700,000 miles and it was 44,213,000 miles from where it leaped from earth to accomplish "a technical feat of the first magnitude." On the 129th day of the voyage—January 2, 1963—its radio became silent; it was then about 54 million miles away, having traveled 223.7 million miles on the deepest successful penetration of the solar system.

Mariner's multimillion-dollar peeps, varying in pitch and tone depending upon the information being transmitted, were music to the ears of scientists, one of whom remarked, "The general effect was as though a young Mozart were practicing a simple etude on a muted harpsichord." They said it might be a long time before the voluminous data—11 million measurements recorded on magnetic tape— could be deciphered; they took to the task of analysis hopeful that the final results would add to man's knowledge of Venus.

A Gold Mine of Data While Mariner rushed on into eternity, scientists jubilantly went to work to interpret the meaning of the eerie signals. Experts at Caltech's Jet Propulsion Laboratory declared that the electronic hieroglyphics from space gave them "a gold mine of data." The magnetometer aboard Mariner revealed that if there is a magnetic field around Venus it is smaller and weaker than the one that surrounds the earth. Also, no evidence was found of an electrically charged belt like the earth's Van Allen radiation belt, and that might be expected because a strong magnetic field is necessary to trap the electrically charged particles. Other observations showed that solar winds blow constantly, and that Venus has a mass 0.81485 that of the earth.

The microwave radiometer indicated a surface temperature of about 800 degrees Fahrenheit, and no evidence of water vapor. Readings on the two infrared wavelengths confirmed it and also showed the temperature on top of the Venusian cloud cover to be about 30 degrees Fahrenheit. Venus, believed to rotate slowly if at all, was summed up as "hot, dry, and dead"; the high temperature ruled out life as it exists on earth.

Dr. Robert Meghreblain, Space Science Chief at the Jet Propulsion Laboratory, described the Mariner 2 experience as "a modest little nursery-school exercise" compared to what will be done in the years ahead. The scientists have their eyes on Mars and Jupiter as goals next on their list. They trust that some new spacecraft's electronic tongue will whisper secrets from the lips of infinity, as Mariner did.

What happened to Mariner 1, launched on July 22, 1962? It was blown up by a range safety officer when, five minutes after take-off, it went erratically off course because an equation related to the ground guidance system was misread by the vehicle's computer. It was explained that the error in the computer equation was merely a missing hyphen that caused the $8-million Atlas Agena B rocket and the $4-million Mariner to be consumed in a yellow-orange fire ball 100 miles above the Atlantic off the Florida coast.

MARINERS TO MARS

Mariner 3, equipped with TV cameras and solar panels, headed toward Mars on November 5, 1964, scheduled to reach its destination in July 1965. Malfunctions "lost" it in space before it had a chance to send back close-up pictures as planned.

Mariner 4, a 575-pound spacecraft carrying TV cameras and scientific devices, was aimed toward Mars, 134 million miles away, on November 28, 1964. It was timetabled to fly by the ruddy planet in mid-July 1965 and transmit television pictures. Not long after the blast-off, tracking equipment indicated it was off course and if it continued in that direction it would pass about 150,000 miles from the target. A radio signal from earth fired a small rocket that lured the craft onto the correct flight path. It was then 1,267,613 miles out in space, traveling 7,019 miles an hour. Thus it returned to the original orientation with an electronic eye focused on the star Canopus, a necessary aim if the TV camera was to be near enough to scan the Martian terrain at the time of fly-by.

After a journey of 228 days, at the appointed hour on July 15, 1965, Mariner kept its rendezvous with Mars—mission accomplished! Its historic voyage was heralded as the most successful and most important experiment man had yet conducted in space, as well as one of the most brilliant engineering and scientific achievements of all time.

While flying within 6,118 miles of Mars, the radio sent back pictures that differed from conventional television; they were transmitted as a series of five million radio signals representing zeros and ones. Those digits were reconstructed by a computer at the Jet Propulsion Laboratory to produce images comprising a series of dots of varying shades, described as somewhat like half-tone photographs in a newspaper. Engineers explained that television creates the image by varying the darkness of each of 525 lines that make up the picture through amplitude modulation. Mariner, however, used a digital or analogue system, less efficient in picture-making but far more reliable when sending pictures across 134 million miles, especially when the transmitter's power is weaker than that used to illuminate the smallest conventional light bulb. Mariner powered by sunlight on its four wing-like panels had only 10 watts of transmitter power.

The series of 21 pictures were acclaimed as "the most remarkable scientific photographs of this age." They showed Mars to be a dead, crater-pocked planet like the moon, no indication of river valleys or ocean basins, and probably the planet never had rain. In addition, data radioed from scientific instruments was interpreted as substantiating the fact that Mars is a desolate place and lifeless. Among the observations were:

1. Virtually no magnetic field and presumably no liquid core.
2. No significant radiation belt.
3. An extremely thin atmosphere.
4. Low air pressure might indicate Mars is swept by winds of unusual intensity.
5. No clear evidence was found relative to "the canals" of Mars, first reported by nineteenth-century astronomers.

Several pictures showed smudges, which were interpreted as possibly fog or haze, or a flaw in the camera system. Therefore, to check the accuracy of the Mars pictures, on August 30—the 276th day of flight—on command from earth ten new pictures of "black space" were recorded on tape and relayed by radio. It was a remarkable feat in communications; the spacecraft was then 11 million miles beyond Mars and 170 million miles from earth.

Distance seemed to lend enchantment. When almost 182 million miles out (September 1965), Mariner's radio continued to send useful signals, 500 hours beyond its designed lifetime. Reports were received relative to magnetic fields and cosmic dust as well as the interplanetary level of cosmic rays and radiation. In its solar orbit ranging from 103 million to 146 million miles from the sun, the spacecraft took 567 days for each circuit around the sun.

When 191 million miles away in October (1965), its radio was shut off by a signal from earth. In early January 1966 the radio was turned on briefly to determine if it was still working; signals came back across 216 million miles. An 85-foot dish antenna in California picked up the signals estimated to be one-billionth of one-trillionth of a watt. The spacecraft was then at its farthest point from the earth in a vast elliptical orbit around the sun.

On its second anniversary—November 28, 1966—Mariner 4 was 206 million miles distant, having traveled more than a billion miles. Three times weekly its 10-watt radioed its condition despite the fact that the signals were rated as "kind of weak"—the strength estimated at one-billionth of a billionth of a watt. Finally, however, its three-year active life ended in December 1967 when a barrage of micrometeors was blamed by scientists at the Jet Propulsion Laboratory for putting it out of commission.

DOES LIFE EXIST ON MARS?

Despite the remarkable performance of Mariner's photographic instruments and the barren, crater-pocked surface pictures, astronomers suggested that negative thoughts regarding life on Mars might be premature. Suppose that some interstellar civilization sailed a spacecraft to within 5,000 miles of the world, its pictures would show no signs of life. Cameras some 500 miles up could not spot objects less than three miles in diameter; they would not "see" buildings, automobiles, airplanes, or even people assembled in crowds. Photographs taken by astronauts prove that to be true, thus substantiating astronomers' conclusions that the early Mariner pictures neither prove nor disprove the possibility of life on Mars.

A NEW LOOK AT VENUS

Mariner 5, a 540-pound instrumented craft raced spaceward on June 13, 1967, scheduled to slide by Venus on October 19 as "the day of encounter," when the planet would be 49 million miles away. No TV camera was on board. While the Mariner was enroute, scientists gained a "scientific bonus" made possible by the longevity of Mariner 4's radio equipment. In August (1967) on its return from Mars to the vicinity of the earth, Mariner 4 was so lined up with Mariner 5 that scientists for the first time could make three-point measurements of electrically charged solar rays. The two spacecraft were 70 million miles apart with the earth roughly right in the middle of a direct line from the sun. Despite the fact that Mariner 4 had been in orbit almost three years and had traveled more than 1.4 billion miles, its radio signals were fairly strong.

True to the timetable Mariner 5 flew around the side of Venus away from the sun, having journeyed about 209 million miles on a long looping path. Speeding at 19,122 miles an hour it swept within 2,480 miles of the planet. When behind Venus and out of radio range, a magnetic tape recorder registered data the instruments detected. Then a tiny radio transmitter played it back to earth as the spacecraft rushed on at increased speed to orbit around the sun. It had about two hours during the fly-by to probe the Venusian atmosphere and to search for radiation and magnetic fields, if any. Venus was so many million miles distant that it took four minutes for the 10-watt signals to reach the big dish antenna at the Jet Propulsion Laboratory. Reception strength was but a whisper, estimated to be no stronger than a billionth of a billionth of a watt. Electronic amplifiers restored the signal strength to readable level.

Analysis of the data led scientists to describe Venus as a hot, inhospitable domain. Based on the radio data conclusions were:

Venusian atmosphere is 7 or 8 times as dense as the earth's, perhaps 15 times, as the Soviet Venus 4 probe indicated.

Venus has no radiation belt such as the Van Allen belt around the earth.

Venus has definite but very weak magnetic activity of undetermined origin.

Venusian atmosphere, which begins at an altitude of 3,800 miles, is comprised of 75 to 85 per cent carbon dioxide; some trace of hydrogen but no detectable oxygen.

Surface temperature was indicated to be at least 500 degrees above zero Fahrenheit.

Many secrets of Venus were still shrouded in its dense layer of clouds: what is the surface like; why is it so hot; of what are the clouds made; how does air circulate around the planet? It was generally agreed, however, that the findings cross Venus off as an inhabited sphere.

TO MARS AGAIN

Mariner 6 embarked on February 24, 1969, on a five-month trip to scan the equatorial regions of Mars. Traveling on a long curving path, the 850-pound craft sailed into the vicinity of the ruddy planet about 60 million miles from the earth in late July. When 771,000 miles from the goal on July 29, transmission began of the first in a series of pictures tuned in at Goldstone, California.

A twin $64-million Mariner 7 took off on March 27, 1969, aimed to swing by the south pole of Mars in early August to photograph the polar "ice cap." These Mariners resembled four-leaf clovers when their 19-foot panels were extended. Sensing instruments were aboard both to search for clues regarding Martian environment—temperature, atmosphere, water vapor, etc., that could support life or even scant vegetation.

Prior to going into solar orbit, the duo transmitted 1,379 pictures of which 143 were far-encounter and 59 were taken from altitudes ranging from 5,700 miles to 2,200 miles. The planet with its vivid South Polar ice cap appeared to be a cold, desolate, crater-pocked globe of jumbled terrain and valleys, and no evidence of life as on earth. Because of the distance it took approximately five minutes for the signals to reach the earth. Conventional TV was not used but instead the digital, or analogue method so effectvely used by Mariner 4.

Acclaimed as the most complex spacecraft ever built in the United States, Surveyor 1 blasted skyward inside the nose of an Atlas-Centaur rocket on May 30, 1966. Seven such flights were planned by similar craft to be designed and built by Hughes Aircraft.

Completing its 247,529-mile lunar journey in 63½ hours on June 2, the three-legged explorer made a gentle landing in the Ocean of Storms from where its radio-television went into action to establish new records in communications. Command signals from the Jet Propulsion Laboratory controlled the flight, and the Goldstone Tracking Station in the Mojave Desert focused its 210-foot dish antenna to pick up lunar TV pictures.

When about 200 miles from the moon an altitude-marking radar switched on and at 52 miles, while the craft was dropping at 5,840 miles an hour, radar ordered the main retrorocket to fire, thereby reducing the speed to 267 m.p.h. at 5½ miles above the surface. The retrorocket comprised more than half the payload; therefore when it was jettisoned, its fuel supply exhausted, the 10-foot-high, aluminum-framed robot weighed 620 pounds. Three vernier rockets applied additional braking thrust until the moon was only 13 feet away. At that point the verniers cut off, leaving the spacecraft to fall at 3 m.p.h., the impact jolt being cushioned by hydraulic shock absorbers and crushable aluminum foot pads.

A half-hour after touchdown, upon radio command, the TV camera began to transmit close-up pictures that won acclaim for Surveyor as an informational technical triumph in electronic communications. The fixed camera televised reflections from a mirror that rotated in all directions and tilted up and down to scan the panorama. Atop the triangular frame was a rotating mast supporting a half-gain antenna and a wing-like panel of 3,960 solar cells capable of producing electrical power of 89 watts. The antenna was designed to transmit TV pictures at the rate of one every 3.6 seconds as the camera looked across the moonscape from a height of 5 feet.

The surrounding topography showed up as a generally flat, hard surface strewn with pebbles, rocks, and boulders, and pocked with craters. The terrain appeared suitable for landing a manned spacecraft. The TV "eye" revealed the robot's feet were on solid ground, not imbedded in dust as some had speculated would be the case. The moon was viewed as "a very gritty, bouldery, pebbly, silt-like place," according to interpreters of the pictures, several hundred of which were received daily for 12 days. After that, however, the frigid two-week night would set in and under such conditions the robot might not survive. The sun was needed for warmth and power. As the long night neared, 400 pictures of the "sunset" were transmitted. Then Surveyor electronically went to sleep. Temperatures dipped to 260 degrees below zero. Because of the moon's slow rotation a day equals 28 days on earth—14 in sunlight, 14 in darkness and extreme lunar cold. Therefore, two weeks would have to pass to determine whether the battery froze. In any event the solar power supply could not be charged during the darkness and hibernation in the "deep freeze."

Chances that the electronic Rip Van Winkle might survive the rigors of the lunar night and intense cold were considered slim. Nevertheless, at the end of the 14-day period the spacecraft amazed its earthly commanders by responding to their signals as the TV camera resumed picture-taking. But on the thirty-sixth day evidence was detected that the remarkable robot was "dying of lunar fever." The battery apparently had suffered some internal abnormality, possibly caused by a short circuit or crack; the internal temperature of the battery climbed to 139 degrees, dangerously near the 140-degree mark, which would be fatal.

Fortunately, the battery temperature dropped and for several days after the "dawn" three stations in California, Australia, and South Africa collected 741 pictures, including photographs of a shattered glass mirror on the box that contained the ailing silver-zinc battery. The box was insulated to maintain temperatures between 40 and 125 degrees around the instruments. A sudden drop in voltage, however, plus another two-week lunar night relegated Surveyor as a silent sentinel on the moon—nevertheless, a monumental signpost in communications.

Suddenly, however, after 86 days of silence, in response to commands from an antenna at Johannesburg, South Africa, the spacecraft again peeped radio signals that were picked up at the tracking station at Canberra, Australia. Surprisingly the battery still had life, having survived 128 days of extreme temperatures from 260 degrees below zero to 500 degrees heat. But the power was too weak to activate the camera.

During its active period Surveyor responded to more than 100,000 commands—256 different kinds—and delivered 11,500 pictures. Said a project scientist, "It just sits there and does what it is told." Appropriately it was proclaimed a signal success.

"Overnight the eyes of Surveyor 1 have become the eyes of the world on the moon," observed President Johnson. "Another exciting chapter in the peaceful exploration of the universe is open for men to read and share."

Surveyor 2 was not as lucky. Departing from earth on September 20, 1966, aimed to land in the Central Bay region of the moon, it got off to a good start, but a steering rocket malfunctioned as it approached the target, causing a crash landing.

PROSPECTING THE MOON

Headlined as America's "first robot space prospector," Surveyor 3 began a trip to the moon on April 17, 1967. As an innovation it was equipped with a steel-tipped aluminum claw about the size of a man's hand mounted on an accordion-like arm that could scoop into the lunar crust. Landing in Oceanus Procellarum only 2.4 miles from the intended target, the rotating camera went into action.

Its photographic prowess, however, was not confined to the moonscape. Peering far into space it photographed Venus. And another Space Age first was recorded when spectacular pictures of a solar eclipse were transmitted. The sun was hidden behind the earth whereas during a solar eclipse as seen on earth the sun is blotted out briefly by the moon. While the spacecraft was in a two-hour shadow cast by the eclipse, the lunar mid-day temperature dropped from 250° to minus 150°.

Among the 6,319 pictures were views of Surveyor itself nestled about 7 feet down the inside slope of a crater that appeared to be 50 feet deep and about 650 feet across. The pictures showed a color-coded wheel designed to help determine the approximate color of the soil. It was attached to one of the robot's legs. The TV "eye" also focused on the facile claw of the miniature "bulldozer" as it gouged trenches 10 to 15 inches long, the deepest 7½ inches. Operated by three electric motors under radio command from technicians 250,000 miles away, the "shovel's" residue was scattered on the craft's white footpad. Analysis of the pictures enabled scientists to appraise the soil as granular "like coarse beach sand after the tide goes out." That was believed to be firm enough to support astronauts. One project engineer described the digging operation and placement of the soil "just where we wanted it," as accurate within half an inch, truly a remarkable feat in communications across a distance of a quarter of a million miles!

Surveyor 4, similar in design to No. 3, except that it was equipped with a magnet to detect magnetic particles, took off on July 14, 1967, aimed to land in the Central Bay region of the moon. It had a rotating TV camera to keep an eye on whatever the magnet picked up and what its miniature digger might scoop from the lunar crust. All went well on the flight until three minutes before the scheduled landing when radio contact was lost. Speculation suggested that the craft was destroyed in a retrorocket explosion during the slowing-down phase prior to impact.

WHAT IS THE MOON MADE OF?

Scientists hoped Surveyor 5, which left the launch pad on September 8, 1967, would send back answers to the age-old question: what is the moon made of? The 616-pound spacecraft landed on a slope inside a crater on the Sea of Tranquility from which the rotating TV camera took more than 18,000 clear pictures of the surroundings and the sun's corona.

An innovation aboard this Surveyor was a 6-inch square gold-plated metal box which on radio command from earth was lowered to the moon's surface on a nylon cord. As a radiation device it was designed to bombard the moon with atomic particles that could help to determine the chemical composition of the soil. Alpha particles from six radioactive sources in the box bombarded a 4-inch square patch. Sensors inside the box registered the number of rebounding Alpha particles, and radio-relayed the measurements to earth.

Scientists studied the signals for clues that might reveal compounds and elements in the lunar terrain. Their analysis concluded that the moon's surface is probably comprised of basaltic soil and rock of volcanic origin as exists "almost any place on earth"—from Iceland to the Hudson Palisades to the Hawaiian Islands and India. In fact, the findings seemed to lend support to the theory that the moon was wrenched from the earth millions of years ago.

Further analysis of the radiation data was said to indicate that the lunar surface consists of 53 to 63 per cent oxygen, 15.5 to 21.5 per cent silicon, 10 to 16 per cent of the elements from phosphorous through nickel, 4.5 to 8.5 per cent aluminum. Anthony Turkevich, of the University of Chicago, who directed the radiation-analysis experiment, said: "The general pattern which emerges is that most of the abundant elements on the lunar surface are the same as the most abundant elements making up the surface of the earth."

The rugged area known as Sinus Medii–Central Bay near the center of the moon's visible face was selected as the destination of Surveyor 6 that took off on November 7, 1967. Sixty-five hours later the 616-pound artifact soft-landed; in a matter of minutes the TV camera began to confirm the findings of previous Surveyors and indicated that the soil is "a kind of sandy loam." Scientists commended the pictures as "the most dramatic close ups yet of the moon"; in fact, they were so clear that the footprints of the robot were telecast as the TV eye scrutinized the terra firma. Also on board was a soil analyzer similar to the one carried by Surveyor 5 designed to help determine the chemical make-up and consistency of lunar soil.

Eight days after landing, on a signal from earth, the robot fired its three vernier engines that made it "hop" up 10 feet in a 6½-second flight and drop 8 feet from the original position, thus becoming the first earth vehicle to accomplish a change of location on the lunar surface; the first rocket-powered take-off on the moon. From the initial landing site more than 12,760 pictures were received at the Jet Propulsion Laboratory, and within a day after the "hop" some 4,000 photos came from the new camera position. A week later the usual 14 days of darkness set in giving Surveyor a rest, after having transmitted 30,027 pictures. When the "reveille" command was flashed the robot failed to respond.

As the finale in the $350-million Surveyor series, No. 7, launched on January 7, 1968, was aimed at the Tycho area of the moon, where it landed on the ninth. The TV camera began sending pictures that showed the rugged, boulder-strewn terrain as scanned near the rim of the huge crater Tycho. Before it ceased operation on February 21 more than 21,000 pictures were relayed, bringing the total for all Surveyors to about 90,000. A steel-tipped claw on a 5-foot aluminum arm furrowed the soil and an "analyzer box" made chemical tests—all operated by radio control from the earth.

But that was not all. A new feat in communications was achieved when a small mirror on the camera picked up two laser beams flashed from Kitt Peak National Observatory in Arizona and Table Mountain Observatory in California. Scientists acclaimed the test "one of the most significant in communications" as the first use of light to communicate over such a long distance—about 240,000 miles. They calculated that the moon's distance now could be measured to an accuracy of six inches by precision light beams that surpass radar in accuracy. The beams, powered by 3 to 4 watts, were about 2 inches wide at the earth and at the moon fanned out several miles, whereas ordinary light would be dissipated by the atmosphere and would be scattered by moisture and dust. The experiment was heralded as demonstrating the feasibility of a really revolutionary technology that well might lead to a myriad of applications. For instance, lunar astronauts would mount reflectors on their moonship so that its location could be pinpointed by laser beams focused on them from earth.

LUNAR ORBITERS

Surveillance of the moon was not confined to surface scanning or surveying. Lunar Orbiter 1, an 850-pound photographic laboratory, was sent up by NASA on August 10, 1966, to take a look at the lunar sphere from on high. As the first United States spacecraft to fly around the moon, it was equipped with a darkroom for picture processing and instruments that converted pictures into electronic signals for transmission to earth.

When about 550 miles above the lunar surface, a rocket fired a braking burst that permitted the four-leaf-clover-shaped craft to be ensnared by the moon's gravity. At that point a complete orbit was made every 3 hours, 37 minutes and 45 seconds. While flying 133 miles above the moon on its twenty-sixth revolution Orbiter was signaled to begin taking pictures. After a number were taped it was maneuvered into a lower flight path about 36 miles high so that the cameras might scan a 3,000-mile strip as potential landing sites for astronauts. It was then looping the moon every 3 hours and 20 minutes in an eccentric orbit that took it out as far as 1,148 miles and at times as close as 26 miles above the moon. On one of the encirclements it transmitted the first United States photo of the far side of the moon. Establishing another first, it photographed and radioed a picture of the earth in a crescent-shaped "last quarter" as viewed from approximately 240,000 miles.

Launched from Cape Kennedy the flight was tracked and maneuvered by the Jet Propulsion Laboratory. Although the Orbiter would probably travel around the moon for about nine months its photographic mission was geared to terminate in about two weeks. The final pictures were snapped on orbits 56 and 57, after which it took up the task of transmitting the images on film the cameras exposed. One astronomer estimated that more lunar data had been returned in seven days than had been gathered from earth-based observations and mathematical calculations in 50 years.

The Orbiter's photographs, appraised as "esthetic and scientific masterpieces," showed 150,000 square miles of the moon's near side and 2 million square miles of the far side. In addition, instruments measured radiation levels and helped to determine the moon's gravitational characteristics.

After a month of picture taking the mission ended, but the spacecraft continued to encircle the moon in an orbit ranging from 220 to 1,023 miles. To prevent its radio from interfering with transmissions from Orbiter 2, NASA decided to crash Orbiter 1. As it flew 1,120 miles above the lunar terrain at 2,161 miles an hour, on October 29 it responded to a command signal that fired a small rocket for 97 seconds. That reduced the speed and started a downward plunge to the moon's hidden side.

Shortly thereafter—November 6, 1966 Orbiter 2 took off on a moon-encircling assignment to conduct experiments in converging stereoscopic photography, a technique that features overlapping photographs of given areas from two different orbits. The flight path ranged from about 25 miles to a high point of 1,153 miles. As the winged electronic-photographic laboratory circled every 3 hours and

28.5 minutes, spectacular pictures were recorded on film and automatically processed for transmission. One widely reproduced scene
showed the inside of the crater Copernicus from a perspective un-
obtainable from the earth. Eleven thousand solar cells covering four
panels exposed to sunlight generated 375 watts to operate the electronic components and to charge a 20-cell, nickel-cadmium storage
battery designed to supply power when the craft was in shadow.

While flying 28.4 miles above the hummocky jagged mountain
terrain the moonscape close-ups featured one view heralded as "one
of the great pictures of the century." The Fauth crater, 13 miles wide
and 4,500 feet deep, showed up exceedingly clear as did Copernicus,
60 miles in diameter and 2 miles deep. Scientists described it as "a
selenographic analogy to the Grand Canyon of the Colorado." They
also noted a striking similarity to the barren Badlands of South
Dakota, or the western slope of the Rocky Mountains in Utah.
Photographic proof now was available that "the moon has had a
long, complicated history of volcanic activity," and that its numerous
craters were not caused solely by meteoric bombardments.

Orbiter 3, similar in design to No. 2, went into space on February
4, 1967. When 242,000 miles from the earth and traveling 4,300
miles an hour, a radio signal ignited a retrorocket to put the craft into
the proper elliptical orbit. Before an eight-day picture assignment
began on February 15 the retrorocket fired again to achieve an
altitude within 28 miles of the moon at the lowest point. The cameras
went into action at 33½ miles and sent back excellent pictures of the
Sea of Tranquility as well as unprecedented views of the craters
Hyginus and Kepler in the Ocean of Storms. The telephoto lens
scanned 2,200 square miles and 11,500 square miles of wide-angle
coverage. Seven months later on September 6, Orbiter 3 was nudged
into a new orbit from 89 to 196 miles above the moon. Finally, with
fuel of the control engines nearly exhausted, Orbiters 2 and 3 were
crash-landed by the space agency on October 11, 1967, in order to
free their radio frequencies for future use.

News headlines on May 4, 1967, reported Orbiter 4 on the way to
film the moon. Maneuvered into an egg-shaped orbit that varied from
1,623 to 3,844 miles above the moon, it took long-shot telephoto and
wide-angle pictures of about 99 per cent of the lunar surface that
faces the earth. And in conjunction with its three predecessors Lunar

4 increased by more than 75 per cent the pictorial coverage of the hidden side. NASA reported that the photos, at least ten times better in resolution than the best existing telescopic views, revealed many geological details previously unknown.

The South-polar area, photographed for the first time from an altitude of 1,856 miles, showed the lunar Antarctica as a rugged terrain with high-rimmed craters and tall ridges. An outstanding picture from an altitude of 1,690 miles featured the Orientale Basin 600 miles in diameter and ringed by the 20,000-foot-high Cordillera Mountains. Geologists are led to believe that the giant basin was formed by the impact of a huge meteorite or comet, possibly 500 million years ago!

A CELESTIAL ART FORM

The fifth and final Lunar Orbiter in the $209-million series, bolted skyward by an Atlas-Agena rocket on August 1, 1967, went into an egg-shaped orbit within 125 miles of the moon at the closest point and 3,760 miles at the farthest.

Designed and built by the Boeing Company, as were all the Orbiters, Lunar 5 weighed in on the launch pad at 850 pounds, including 150 pounds of photographic equipment and 260 pounds of engine propellant. Again the Eastman Kodak photographic unit was used, comprising elements for making photographs, film processing, and converting the image on film into electrical analogue signals radioed to earth via a communications system developed by the RCA Astro-Electronics Division.

The craft was 5 feet high and 5 feet wide. Once in space two antennas and four solar cell panels unfolded; the antenna booms extending on opposite sides spanned 18 feet from tip to tip while the solar panels spanned 12 feet.

The signals received at NASA tracking stations in California, Australia, and Spain were converted into light by a kinescope and the images were then transferred to 35 mm. film. Pictures of the moon's face and hidden side were of excellent quality. From 3,640 miles above the orb, a spectacular picture was taken of the earth as seen from a distance of 224,000 miles. The telephoto lens brought about five-sixths of the globe into view. Easily identified were the African east coast, Italy, Greece, Turkey, the Red Sea, the Arabian Peninsula, the Suez Canal, and India.

The Orbiters remained in orbit for some time after completion of their photographic missions, totaling 1,950 pictures. They continued to radio data on radiation, micrometeoroids, and gravitation. Eventually each craft, except No. 4, was deliberately crashed on the moon to avoid any possible interference with communication by future spacecraft. The fourth Orbiter was believed to have crashed in October 1967, after having been listed as lost in July. Nevertheless, it succeeded in photographing the front face of the moon. By the time Orbiter 5 transmitted pictures of the near and far sides, scientists had an album of the entire façade of the moon and 99.5 per cent of the opposite side. Celestial photography was acclaimed a new art form.

Supplementing the Orbiter lenses, another satellite Dodge (Department Of Defense Gravity Experiment), equipped with TV cameras, was credited with the first color photographs of the full-face view of the earth transmitted from an 18,100-mile-high orbit. The pictures released by the Defense Department showed a hurricane approaching the Texas-Mexican border and two other swirling storms churning above the Atlantic. The recorded photos radioed to earth in October 1967 were taken in July and September.

LUNAS

The year 1959 was opened by the Soviet Union with the introduction on January 2 of a series of Luna (Lunik) satellites for lunar probes. Luna 1 was believed to have passed within 3,730 miles of the target and then went into orbit around the sun.

Luna 2, orbited on September 12, 1959, was reported to have crashed on the moon ending its radio transmissions after a 35-hour flight. Luna 3 launched on October 4 looped around the moon and radioed the first pictures of the far side as it swung into a wide orbit that eventually led to destruction in the earth's atmosphere.

A 135-pound "automatic station" known as Luna 4 stirred speculation on April 2, 1963, that the Russians were aiming to land a radio-instrument package. Maneuvering seemed to have failed and it was reported passing about 5,000 miles from the lunar target before going into a solar orbit.

The Soviets took another shot at the moon on May 10, 1965, when Luna 5 was launched equipped with radio and other scientific instruments. The plan was reported to be achievement of a soft landing so that the intact apparatus could function from the lunar surface. A malfunction of the braking rockets which were supposed to slow the craft on its approach from a velocity of 6,000 miles an hour to about 6 miles an hour at touchdown was said to have resulted in a crash instead of a gentle landing that might have preserved the instruments.

A month later—on June 8—Luna 6 headed for the moon. An engine used to maneuver the vehicle failed to shut off and was blamed for a miss of the target by 100,000 miles as Luna sailed off into outer space.

Luna 7's fate was not much different as far as accomplishment of its mission. It was a 3,000-pound spacecraft believed to have been aimed at the moon. Launched on October 4, 1965, it crashed at impact on the eighth. The Soviets announced, however, that it carried out "most of the operations necessary for a soft landing during the approach to the moon."

Luna 8, also unmanned, went skyward on December 3, 1965. Arriving at the moon precisely on schedule it failed to make a soft landing and crashed. All systems were reported functioning normally at all stages of the descent except the final touchdown.

Adhering to the old adage, "If at first you don't succeed, try, try again," on January 31, 1966, the Russians rocketed 3,489-ton Luna 9. It was described as "a crash-proof, uncontrolled photographic capsule jettisoned by its carrier rocket just before impact, allowing it to fall free under the pull of lunar gravity" in the Ocean of Storms.

An intricate coordination of the retrorockets slowed the spacecraft from the approach velocity of 6,000 miles to 6 miles an hour. As soon as the station settled down the antennas automatically opened and transmissions of telemetric signals including television began on 183.538 megacycles. For the first time in history communication was established with another heavenly body—a radio bridge from moon to earth. The feat was heralded as marking the beginning of on-the-surface exploration of the solar system. A new dimension had been added to man's capability for observing and exploring the universe.

Ten panoramic television pictures described as sensational, amazing, and fantastic were picked up by the 250-foot dish antenna at

Britain's Jodrell Bank radio observatory. Indeed the British gained a "scoop" by being first to release the moonscapes which showed the rock-strewn crust pockmarked with craters. Two days after the landing—February 5—Soviet space officials announced that Luna 9 in 48 hours of battery-powered lunar life had completed its research and photographic mission; they boasted that "the moon speaks Russian."

Luna 10 lifted off on March 31, 1966, and went into orbit around the moon on April 3 "to carry out a broad study of the physical characteristics of the moon and near-lunar space." One instrument—a gamma ray spectrometer—was designed to make critical measurements of the nature of the terrain. Another device could measure the effect, if any, as the moon passed through the "comet tail" of the earth. Tests of infrared emissions also were on the agenda. A meteorite particle recorder was on board as well as a device to probe for magnetism as the "laboratory" circled the orb at a maximum distance of 620 miles, minimum 220. Apparently no picture taking was attempted.

Designed "for testing of systems of an artificial moon satellite and scientific exploration of near-lunar space," Luna 11 left the launch pad on August 24, 1966. Once in orbit the 3,608-pound craft was reported circling the moon every 2 hours, 58 minutes at altitudes ranging from 99 to 745 miles. Evidently there was no picture transmission.

The ninth spacecraft to scout the moon with a camera was Luna 12, lofted on October 22, 1966. Described as a "research craft" it was to orbit, not land. Pictures were clicked off for Moscow television as the camera scanned within 58 miles of the moon in an orbit that extended as far as 940 miles.

Luna 13, an automatic research station, was launched on December 21, 1966, the same day that Cosmos 137 went into space. After an 80-hour flight the Luna soft-landed near the Ocean of Storms, from where television scenes of the moonscape were transmitted while other instruments registered radiation and measured surface density as well as firmness. The information radioed to earth was interpreted to indicate that to a depth of 8 to 12 inches the lunar surface is quite similar to that of the earth.

Luna 14 launched on April 7, 1968, became the fourth Soviet

lunar orbital research station. It was intended to circumvent the moon in an orbit ranging from 540 miles to a low point of about 100 miles, completing a revolution each 2 hours and 40 minutes.

The headline SOVIET LAUNCHES UNMANNED CRAFT TOWARD THE MOON on July 12, 1969, led to rumors that an attempt was under way to land on the lunar terrain, scoop up samples of soil, and return to earth. Russian reports said Luna 15 was for "scientific research" and for testing new automatic navigation systems. After 52 revolutions during which it was reported within 10 miles of the surface, the exploit ended on July 21. Radio silence led to the belief that the craft crash-landed in the Sea of Crises.

Mars 1 A 1-ton space vehicle was hurled upward by the Soviet Union on November 2, 1962, scheduled for a seven-month, 60-million-mile trip toward Mars. By June 1963, it was scheduled to be near enough to the ruddy planet to take pictures and radio them to earth; the radio operated on 922.76 and 183.6 megacycles. It was hoped that the probe might reveal whether there is life on Mars; and in any event the foray might establish "super long-distance radio communication." When the satellite was 106 million kilometers from the earth, on April 21, 1963, radio contact was lost, attributed to a possible defect in the antenna of the orientation system.

When Mars 1 was 160,000 miles from earth it was reported to have been photographed with the aid of a 100-inch reflector telescope at the Crimean Astrophysical Observatory. It was believed to be the first time an interplanetary vehicle was photographed over such a distance.

Flight 1 (Polyot 1) A maneuverable satellite—an unmanned earth-controlled spaceship—was launched on November 1, 1963, by the Soviet Union as an advance toward perfecting rendezvous and dock of two or more vehicles in space; also military and weather satellites could someday be maneuvered and spacecraft steered away from radiation zones. Obeying radio commands from earth, the craft was reported to have turned from one side to the other, soared, dived, and shifted from an initial almost circular orbit with an apogee of 368 miles to an elongated orbit with an apogee of 892 miles. A day or two later, when the fuel necessary for maneuvering apparently was exhausted, the ship went into a "final," or fixed, orbit around the earth.

Polyot 2, orbited on April 12, 1964, was described as a steerable space vehicle "necessary to assemble powerful ships in space for interplanetary expeditions." Apogee was reported at 311 miles and perigee 193 miles from the earth.

LOOKING IN ON VENUS

In its first attempt to investigate the planet Venus, the Soviet Union sent a satellite Venera 1 aloft in 1961. Radio contact was lost a month after the launch.

Venera 2, a 2,123-pound instrumented vehicle, was orbited on November 12, 1965, bound on a 3½-month, 180-million-mile trajectory. While it was 715,000 miles out in space—on November 16—Venera 3, a 1-ton craft described as "somewhat different in design in the make-up of its scientific instruments," was sent up "to solve a number of new scientific tasks."

When the cloud-enshrouded planet was 38 million miles from the earth on February 27, 1966, Venera 2 passed within 25,000 miles, too far for its photographic mission to succeed.

Venera 3, after a 106-day flight, was believed to have crash-landed on the target on March 1, 1966.

What was to have been Venera 4, launched on November 23, 1965, was reported to have exploded while still in earth orbit. Apparently to disguise its purpose it was referred to as one of the Cosmos series (No. 96). Therefore, the next shot on June 12, 1967, was designated Venera 4, a 2,437-pound, cameraless, "automatic research station" scheduled to reach the goal after a four-month, semi-circular trajectory of 217.5 million miles. Its mission: to gather information about radiation, temperature, magnetic fields, atmospheric density, pressure, etc., and radio the findings to earth.

When the vehicle approached the target on October 18 an instrumented capsule automatically parachuted toward the surface with two thermometers, a barometric pressure sensor, an atmospheric density gauge, and several gas analyzers. During the descent, estimated to have reached about 15 miles above the terrain, data was radioed for 90 minutes. Silence after that was attributed to the heavy atmospheric pressure and torrid heat that destroyed the capsule and prevented a soft landing of either the Venera or its package of instruments.

Analysis of the signals was said to reveal that the murky Venusian atmosphere is 15 to 20 times as dense as on earth, and consists almost exclusively of carbon dioxide, about 1.5 per cent hydrogen, 0.4 per cent oxygen, 1.6 per cent water vapor. No trace was found of nitrogen, which makes up 76 per cent of the earth's atmosphere. Temperature was indicated to range from 104 to 536 degrees Fahrenheit. No magnetic field or radiation belt was reported in evidence.

Based upon data from Venera 4, coupled with findings of the Mariner spacecraft and powerful American radar observations, scientists at the Jet Propulsion Laboratory determined Venus to be much hotter and far more crushing in pressure than the Soviet measurements indicated. On the surface the temperature is believed to be about 900° F, and the atmospheric pressure 75 to 100 times that at sea level on earth. Also, American analysis of all available data established the planet's radius to be about 3,759 miles.

Seeking additional information and to confirm previous data relative to Venus, a 2,491-pound Soviet spacecraft Venera 5 took off on January 5, 1969, on a 217-million-mile, 130-day flight that ended on May 16. An instrument-laden capsule was parachuted, and for 53 minutes during the descent measurements were made of the temperature, pressure, and chemical composition of the atmosphere.

Venera 6, which departed on a dual mission five days after No. 5 was launched, sailed into the clouds of Venus on May 18 about 180 miles from the point where her sister spacecraft entered. Again an instrument capsule parachuted, and according to reports from Moscow "fulfilled its mission" by transmitting data for 51 minutes during the drop. Both vehicles entered the night side of the planet to avoid solar interference with radio. Signals from Venera 5 were reported heard during 22 miles of the descent; from Venera 6 for 24 miles.

Interpretation of the data released from Moscow led to the surmise that the extreme atmospheric pressure might have crushed the capsules and silenced the radio. Temperature was indicated to range from 750 to 1,000 degrees; atmosphere comprised of about 95 per cent carbon dioxide; oxygen less than $\frac{4}{10}$ of 1 per cent. Both flights were classified as flawless, and indeed were remarkable feats in electronic navigation across 217 million miles to reach such a distant target on schedule with pinpoint accuracy.

XIII   Twins of Science:
Maser and Laser

> Approach, thou beacon to this under
> globe,
> That by thy comfortable beams
> . . . Nothing almost sees miracles.
>   —*King Lear*

THE MASER

CHARLES H. TOWNES,[1] PHYSICIST AT COLUMBIA UNIVERSITY, FROM
theoretical speculations regarding microwaves created one of the
most revolutionary devices of the age—the maser (pronounced *may-
zer*).

Basically, the maser evolved from discoveries made about 60 years
ago by Albert Einstein and Max Planck, a German physicist. They
observed that atoms possess varying amounts of energy—some atoms
have relatively high energy levels and others low. They also noticed
that the energy status of certain atoms could be altered by outside
signals. For example, a low-energy atom might absorb the signal and
turn into a high-energy atom, and a high-energy atom might have
some of its energy knocked off and become a low-energy atom.

Mindful of these concepts, Dr. Townes in 1954, seeking a way to
extend the range of microwaves, reasoned that if materials could be

---

[1] Dr. Townes went from Bell Telephone Laboratories to Columbia University
in 1948 as associate professor of physics; professor of physics 1950–1961; vice
president and research director Institute of Defense Analysis, Washington,
D.C., 1959–1961; provost and professor of physics Massachusetts Institute of
Technology 1961–1966; recipient of the Nobel Prize for Physics 1964; professor
at large, University of California, 1967.

found containing large numbers of high-energy atoms vibrating at certain frequencies—and if these atoms could be struck by microwaves of the same frequencies—then the energy given off by the atoms would tremendously strengthen the microwaves. He directed microwaves into ammonia gas, which is packed with high-energy atoms, and when the microwaves emerged from the gas they were greatly amplified. From that process—Microwave Amplification by Stimulated Emission of Radiation—the maser got its name.

Subsequently, in 1957, Nicolaas Bloembergen, Professor of Applied Physics at Harvard University, proposed a practical way to build a solid-state maser as a low-noise microwave amplifier. The first maser of this type, built at Bell Laboratories, featured a synthetic ruby crystal,[2] which to operate effectively had to be cooled to minus 456° F. (by jacketing it in liquid helium).

A ruby amplifier was chosen because it is uniquely free of noises that interfere with radio signals. For example, it does not have the hot cathode or hurtling electrons that generate noise in conventional amplifiers. In fact, the ruby amplifier is so quiet that only noise made by matter itself in heat vibrations remains; but at temperatures close to absolute zero, even this is silenced. Thus, very faint signals from satellites are uncontaminated and amplified with clarity.

Indicative of the maser's power of microwave amplification is one scientist's report that by using what he termed a relatively crude maser, instead of other amplifiers, he has picked up radio signals from stars three times as far away. It was such a maser at the Naval Research Laboratory that facilitated measuring the waves of energy that radiated from Jupiter, and calculating the surface temperature of that planet as minus 150° F.

TYPES OF MASERS

At the outset there were four main types of masers: (1) Gaseous, using ammonia for microwave amplification operated in pulses. (2) Solid-state, which used a ruby or several other types of crystals for

2 A ruby is a single crystal of aluminum oxide with a small fraction of the aluminum atoms replaced by chromium atoms, the source of its red fluorescence. Machined in the form of rods, which may vary in size, the ruby possesses an extraordinary property when chilled and subjected to a magnetic field; it can be excited to store energy at microwave frequencies. As the signal passes through a ruby, it releases energy and the signal is amplified more than 4,000 times.

microwave amplification operated in pulses. (3) Solid-state optical maser, using a ruby or other crystal for infrared amplification operated in pulses. (4) Gaseous optical, using a mixture of helium and neon for amplification of light, operated continuously instead of in pulses and held promise for practical applications in communications.

The Optical Maser In 1958, Townes and Arthur L. Schawlow extended the maser principle from the microwave region to the infrared and visible light spectrum. Since the objective is to amplify light, such masers are called optical masers or lasers (*layzer*).

Laser is an acronym for Light Amplification by Stimulated Emission of Radiation. It is an instrument to generate, amplify, and control light waves; it is to light what the electron tube is to electronics. From laser a new verb has emerged—to lase—referring to coherent light produced by laser action. Thus of a laser, or of an active material, it may be said that it lases, it is lasing, or it lased.

The first theoretical analysis that showed it should be possible to build an optical maser was published in 1958 by Townes and Schawlow.[3] It remained for Theodore H. Maiman, then of Hughes Research Laboratories and later president of Korad Corporation, to discover that a single crystal of pink ruby by addition of about 0.05 per cent chromium was a suitable active component for a laser. He announced it in July 1960, and is credited as the first to observe optical maser effects in ruby.

The heart of the first optical maser developed at Bell Laboratories was a synthetic ruby rod, said to have been produced originally for such use by Schawlow and used in a manner originated by Maiman.

All of the early optical masers operated in short bursts, or pulses, less than a millionth of a second in duration. In that split second, however, the ruby had a chance to cool. In continuous operation terrific heat burned it out; a cold method became the goal. It was achieved at Bell Laboratories by continuous operation of solid-state optical masers in three solid materials including ruby. This led to the prediction that the power of a few thousandths of a watt would be increased and the laser's range of application extended.

[3] Basic concepts of the optical maser were disclosed by Schawlow and Townes in *Physical Review,* December 1958, and in U.S. Patent No. 2,929,922. (Schawlow was a staff member of the Physical Research Department of Bell Laboratories, which he joined in 1951; in 1962 he became Professor of Physics at Stanford University.)

Scientists were convinced it would be feasible to develop a communication system by modulating the output of an optical maser as a radio wave is modulated to carry voice, music, and television. Indeed, the laser is appraised as foreshadowing expanded use of light as a new medium of communication, the possibilities of which are described as breath-taking. For instance, since the amount of information that can be carried by a laser-beam communication channel is proportional to its frequency, in principle there is room for 80 million TV channels in the visible spectrum between the wavelengths of 4,000 and 7,000 angstrom units (unit for measurement of wavelength, equal to a hundred millionth of a centimeter). And extending out on either side of that waveband are bands of ultraviolet and infrared, which can be explored and exploited by the laser.

It is calculated that the focused beam of a laser could shine as a 1-foot spotlight on a screen 100 miles distant. Because of the nature of light itself, however, laser light beams are absorbed or obstructed by buildings and walls, and by atmospheric factors such as fog, rain, or snow. In attempting to burn through, much of the beam's energy is lost. But in the airless void of outer space such limitations do not exist. And in the form of ultraviolet and infrared light even relatively low-power beams can be modulated to communicate across many miles of clear, open space. For earthbound service engineers are studying the possibility of solid glass fiber waveguides through which laser light beams can be routed, bent, and amplified, and in the process avoid weather interference.

Coherent Light Since light is an electromagnetic wave, it has many characteristics in common with other waves and can be thought of in terms of a procession of crests and troughs, like waves of water. Light from a conventional light source is *incoherent,* like the scramble of tiny separate waves stirred up by throwing a handful of pebbles into a pool. In contrast, however, if a single pebble is tossed into the pool a *coherent* circular wave front is created. A salient feature of laser-generated light is that the waves are all in step—coherent; they are monochromatic (close to one wavelength or frequency), whereas incandescent or fluorescent light from an ordinary light source is nonmonochromatic. Laser light is an extremely fine radiation of wavelengths which from crest to crest are measured in microns, a few tens of millionths of an inch!

Communication experts explain that light waves vibrate at frequencies tens of millions of times faster than radio waves, and because of this, a beam of coherent light has exciting potentialities for handling enormous amounts of information—telephone calls, TV programs, data messages, etc. Putting it simply, ordinary light waves from an electric lamp move "like an unruly mob," while coherent light waves move "like disciplined ranks of soldiers," which can be maneuvered like radio waves.

"The optical maser produces an intense and extremely narrow beam of light," reports C. G. B. Garrett of the Bell Telephone Laboratories.[4] "Within its narrow cone and frequency band, the beam is more than a million times brighter than the sun. The light generated by the laser is coherent; that is, light in which a definite phase relationship exists from point to point in all parts of the beam. Since it is the property of coherence in radio waves which makes it possible to control, direct, and modulate them, it may also be possible to use coherent light in the same way."

So promising and encompassing are the applications that scientists declare the effect in multiple fields cannot even be calculated yet. Over the next five or ten years they expect new and revolutionary developments as masers, including the laser, are improved and advanced in usefulness. Harvard University already has two maser (microwave) clocks believed to be accurate within a second in a million years; oscillations of hydrogen atoms serve as a frequency standard.

HOW THE LASER WORKS

A solid optically pumped laser features a ruby rod about 6 inches long, around which is coiled a powerful electric flash lamp. Intense bursts of light excite or pump chromium atoms interspersed in the ruby crystal to a high-energy state. And it is the presence of chromium that gives the ruby its pink, or reddish, hue. Yellow, green, and ultraviolet light from the flash lamp are absorbed by the ruby, causing the chromium atoms to emit red light. Both ends of the ruby rod are polished to act like mirrors. As a red ray aims at either end of

[4] *Electrical Engineering,* April 1961.

the rod it reflects back into the crystal, bouncing back and forth between the two mirrors. In the process other chromium atoms are hit, stimulating them to produce more red rays, all coherent, parallel, and essentially a single wave. This chain reaction within billionths of a second creates a powerful beam that bursts out one end of the rod, which is deliberately made more transparent than the other.

A gas laser (helium-neon) with the protective cover removed looks like a long neon sign lamp with an adjustable mirror at each end. Excited by a radio frequency, or direct current, the output is a continuous-wave, pencil-thin beam the power of which is rated, at "best," to be about 0.15 watts; but no doubt will be pushed to higher wattage. The wavelengths produced range from a deep orange through several different wavelengths to the middle infrared, where wavelengths are about 20 microns. The visible light beams can be seen projecting from the end of the device.

Semiconductor injection-type lasers feature a different mechanism than the gas and crystal lasers. For example, a tiny crystal of gallium arsenide is formed into a structure not unlike a junction transistor with atoms lacking electrons (called "holes") in one area and atoms with excess electrons in the other. This is achieved by "doping" the crystal with an additive; one area is treated to give it more free electrons than the other. Both sides, or reflecting surfaces, are polished optically flat. When an electric current is injected across the crystal the atoms are excited to a high-energy state, causing the electron holes to migrate across the junction, which is only one ten-thousandth of an inch thick. There in the junction zone, the electrons "drop" into the holes and surrender excess energy as light that bounces back and forth, finally emerging from one edge of the laser as a tiny amplified beam of coherent infrared light.

Semiconductor junction-diode, or injection, lasers are noted for efficiency, simplicity, and compactness. And their output is easily and quickly controlled by altering the supply voltage so that a signal can be impressed on the light beam.[5] Efficiency in production of light by the semiconductor laser is estimated to be in excess of 50 per cent, while other types are rated at much lower percentages. Thus the semiconductor "family" opens new areas for ingenuity in communi-

[5] "Junction" refers to the structure; "injection," to the exciting process.

cations, computers, and other fields. It is highlighted as a key to development of lightweight laser transmission equipment for communications and remote control, including hand-held, private-line communications systems and ranging devices as well as point-to-point contact with spacecraft. One of the semiconductor's main attributes is elimination of the high intensity light sources for laser pumping; and low voltage requirements widen its range of applications.

Evaluating the advantages gained by the laser, scientists point out that a number of disadvantages remain that must be overcome. But, says a laser expert, "I doubt that the limitations will be permanent; it should be understood, however, that we have a long way to go."

LASER APPLICATIONS

It is certain that the optical maser will extend into various fields unrelated to communications; for instance, metallurgical, medical, chemical, spectroscopic, and many other applications that include welding even of microscopic parts. A laser beam properly focused can not only weld and cut but melt or vaporize. For example, scientists at the Bureau of Mines have found that laser beams may provide a new means for making valuable industrial chemical products from coal. In a split second the high-energy light waves vaporize or disintegrate coal into its chemical components.

Laser microwelding of thin materials to each other and to other materials not ordinarily weldable by conventional methods has been reported by Maser Optics Incorporated of Boston.

Laser microhole punching is another development, with holes punched less than a molecule in diameter. And since the laser's energy can be sharply focused—through a hole 50 millionths of an inch in diameter—scientists envisage the possibility that it could be concentrated within single living cells to perform treatment of deseased tissue which may be called laser-surgery. For example, an instrument known as a laser retina coagulator, developed by physicists at the American Optical Company, has been used by eye specialists at the Columbia-Presbyterian Medical Center to destroy a tumor in the blood vessel system of a patient's retina in one-thousandth of a second. It is possible that a torn, detached, or injured retina can be "welded" to its support by coagulation when the laser beam is applied for a split second.

Surgeons at the Children's Hospital, Cincinnati, Ohio, removed a tumor from a man's thigh by using a laser beam for a knife.[6] Despite the fact that small blood vessels around the tumor were severed during the 15-minute operation, none of them bled because the laser's intense light cauterized as it cut. Doctors are reported to be studying the bloodless method for potential use in operations on the liver, spleen, or even the brain without fear of hemorrhage.

A "light knife" that permits surgeons to use the focused laser beam as easily as they do a scalpel has been developed at Bell Telephone Laboratories. The beam is guided from the laser source through a hollow, jointed arm to a small probe about the size of a fountain pen which is held like a scalpel. The probe also may be attached to a surgical microscope for more delicate operations.

FORWARD STEPS

Laser concepts and developments during 1966–1969, and some of the technical possibilities for future application:

1. Piercing diamond dies with increased efficiency.

2. Alignment purposes in manufacture of precision tools.

3. Speeding pipe laying by assuring correct alignment so that the construction crew need not stop every few minutes to make sure the alignment is right.

4. Laser-guided mining and tunnel digging with extreme accuracy. An alignment laser device has been used to prevent a boring machine from deviating a few fractions of an inch off course in digging a 2-mile irrigational tunnel through a hill. So successful is this straight and narrow beam that it is believed that someday it may replace the old transit and plumb line in alignment and surveying projects.

---

[6] For the operation performed on January 24, 1966, by Dr. Thomas E. Brown, a gas laser known as the argon or "green laser" was used, one reason being that the green light is best absorbed by red objects like blood cells. Also, the green laser beams a steady stream of light and has a power output of up to 4 watts, which is more than that of other continuous lasers. Thus it is more easily controlled at the focal point and generates enough heat to "cut through" tissue.

5. Extended use in bloodless surgery and in the development of a laser microscope.

6. Increased use in data processing and television.

7. Tiny traces of metals in a single human cell, or part of a cell, can be detected by application of the laser. The light beam is focused through a microscope on a minute piece of tissue, which is vaporized at about 18,000 degrees in a few millionths of a second. Light from the puff of smoke consists of many wavelengths that produce a spectrum of the cell's chemical composition. Intensity of the color at each wavelength depends on the quantity of a particular metal present in the sample. Experiments at Stanford University School of Medicine detected metals such as iron and zinc in single red and white blood cells, sperm cells, liver and kidney cells.

8. Micromachining of thin films, for example in thin-film integrated circuit work and machining of delicate circuit patterns; the laser light is focused to a tiny spot for precision cuts less than 5 microns wide and a trimming accuracy to 0.1 per cent.

9. A new experimental technique of recording computer data developed at the Jet Propulsion Laboratory utilizes a highly focused beam of intense light of a laser. Linked with TV cameras it is reported that this method could store 1,000 times as many pictures as the video-tape system used for interplanetary photography. The laser beam is focused on the magnetic film through a microscope and reduced to micron size—a pinpoint to the human eye. Billions of bits of information could be stored in that tiny space. Engineers believe it may lead to a generation of more compact computers in which as much information could be stored on a square inch of film as presently contained in a 10-cubic-foot memory unit.

10. A "gas transport" laser developed by Sylvania with a power output of 1,000 watts.

11. Geophysical laser strain meters in the form of long inferometers operating with laser beams detect with extremely high sensi-

tivity and accuracy distortions or disturbances in the earth's crust, some so weak as to be undetectable by seismographs. A number of industrial and university laboratories are developing and testing such instruments.

Mindful that in FM radio the frequency rather than the intensity (as in AM radio) is varied, or modulated, Dr. William J. Thaler, physicist at Georgetown University, applied for a patent on a device to communicate over a laser beam by varying the beam's frequency instead of the intensity. He calculated that by applying this technique, interference from extraneous light sources would be avoided in much the same way that man-made noises and static are minimized in radio by the use of FM.

While lasers are foreseen to have immense practical application on earth, they will also be of great value in seeking out the wonders of the universe through satellite and space-probe communications; they show promise of revolutionizing radar and greatly multiplying channels of communication.

Little wonder that scientists, researchers, engineers, and industrialists are excited about the laser. For satellite communications, for signaling between space vehicles and interstellar exploration, this quantum-electronic device arrived just at the right time.

A light beam from a powerful laser was pinpointed on the moon on the night of May 9, 1962, when a team of scientists from Massachusetts Institute of Technology and the Raytheon Company aimed a laser at the lunar orb and hit it with a beam of ruby light. It lit up a spot only 2 miles in diameter, whereas a radar microwave beam would have spread over an area of about 500 miles.

Engineers explained that if a man had been standing in the mountainous area southeast of the crater Albategnius he would have noticed that the dark landscape was lighted by a succession of dim red flashes comparable to the illumination produced on the walls of a room by a flashlight bulb. Proof of the bull's-eye came within 2.6 seconds following each of 13 successive hits, when reflections of the light returning to earth after the round trip of 238,857 miles were registered on an oscilloscope.

A laser operated by Kiyo Tomiyasu, a physicist in the General Electric laboratory, drilled holes in a pea-sized black synthetic diamond. Each hole was drilled by a light flash only two-thousandths of a second in duration. Scientists say that they see no reason why laser light must be confined to red, and some of them are interested in blue-green for underwater exploration probing the ocean floor, shipwrecks, and other obstacles at least a thousand feet away, farther than ordinary light penetrates. And such beams could act as communication channels between submarines.

The laser's intense monochromatic light also makes it an ideal tool for schlieren, or shadow, photography, which permits viewing density changes in a transparent medium.

Another laser use is light-beam radar, said to be 10,000 times more accurate than that achieved by microwaves. Engineers of the Sperry Gyroscope Company state that such a system can measure with absolute precision speeds varying from spaceship orbital injection velocities of 5 miles a second to virtual stop—less than one ten-thousandth of a second. It is known as "laser-doppler radar," operating at a frequency of trillions of cycles per second. It is estimated that such a system could easily detect and measure the movement of a spacecraft edging up to a space station even at a fraction of an inch per second, so that a control system using a signal of such sharp precision would permit a huge vehicle to dock as lightly as a feather.

A sun-activated laser that uses sunlight without converting into another energy medium such as electricity or heat has been developed at RCA Laboratories. Sunlight is directly converted into a continuous beam of coherent infrared light. This opens the possibility of sun-powered lasers on future space vehicles, as intense coherent light rays are used in communications as well as for tracking and geodetic measurements of the earth.

Laser satellite-tracking offers two advantages over radio tracking; it is more accurate and requires no electrical power in the satellite, thus permitting indefinite tracking of a passive or silent satellite. The time required for the laser's light beam to travel to the satellite and back, and the angles of the returned light beams, enable calculations of the satellite's position.

The laser also becomes a new tool of science. It may be used to confirm or disprove old theories. For example, in 1881, A. A. Michelson and E. W. Morley showed the velocity of light to be unaffected by motion of the earth through space. Using a helium-neon gas discharge laser, scientists at M.I.T. repeated the experiment with outstanding accuracy and thus again laid to rest the old theory that light waves are propagated through an all-pervading ether. If that were true the velocity of light would not be constant. Thus the Michelson-Morley experiment discarded the ether idea and set the stage for Einstein's theory of relativity. Lasers also have been used in tests to detect the exact rotation of the earth.

SIGNPOSTS OF PROGRESS

Indicative of advances in development and use of the laser in communication, the U.S. Army in February 1965 announced that the Army Electronics Command laboratories at Fort Monmouth, New Jersey, had succeeded in transmitting all seven of New York City's TV channels simultaneously on a laser light beam.

Significant progress in lasers at RCA from 1965 to 1970 included:

1. A high-powered laser range finder for use in a missile tracking system, providing an accuracy of plus or minus 2 feet at ranges up to 10 miles. Emitting 50-million-watt peak power pulses at a rate of ten per second, it acts like radar, determining the distance of an object by measuring the time required by the pulses of energy to travel to an object and back. The installation site was the Naval Ordnance Test Station, China Lake, California.

2. A supersensitive light detector that can sense up to 100 million intensity changes a second on a beam of light, giving it ability to distinguish as many as 25 separate TV programs being carried simultaneously on a single laser beam. Engineers declare that all the radio and television programs as well as all the telephone conversations in the United States in a single day could be put on one laser light beam.

   It was a major problem to build a receiver sensitive and fast enough to detect and process information. The new device is said to be the first light detector with the sensitivity, speed, and frequency range that could make possible practical laser communications across the optical spectrum. Operating on alternating current it is responsive to the entire range of optical frequencies from infrared through the visible and ultraviolet portions of the spectrum.

3. A compact "two-in-one" solid-state laser that produces high power continuous with high efficiency creates a 10-watt beam of continuous infrared radiation. The combination of high power and efficiency is said to make it extremely adaptable for such tasks as welding, micromachining, drilling, and working refractory materials as well as for short-range radar and line-of-sight communications systems.

4. A hand-held laser beam transmitter about the size of a home-movie camera and weighing 6 pounds was developed for NASA for experiments during the Gemini 7 flight. Designed to transmit 25 watts of light power it was subjected to various tests related to communications to and from spacecraft.

5. The feasibility of using sunlight to power a laser was demonstrated as a step toward a 50-million-mile communication system when a TV picture was transmitted over a narrow light beam for about 6 miles. The sun-pumped laser stems from the need for long-distance communication from spacecraft to earth using as little power as possible. The laser optically pumped by sunlight and collected or concentrated by a parabolic mirror minimizes or eliminates the use of spacecraft power.

6. A new argon gas laser for possible use in data processing, space communications, satellite tracking, earthquake detection and "bloodless surgery" has three times the service life of previous units. Because the laser cauterizes as it cuts, it is believed that surgeons will be able to perform "bloodless" operations on such vital organs as the liver, where bleeding during surgery is a major problem.

7. A neon gas laser that produces intense beams of ultraviolet light continuously up to 1,000 hours has many promising applications in various fields including optical recording on materials insensitive to visible light, in chemical processing, and the photographic as well as copying arts.

8. Linking laser technology and TV, a revolutionary system was developed to send and record images of photographic quality ten times sharper than conventional TV pictures. Designed for use in the Earth Resources Observation Satellite (EROS), the system affords high-resolution photos of the earth and planets, and is seen to have many industrial applications, including microcircuit manufacture, news photo transmission, and graphic arts production.

9. A multi-purpose compact laser system can serve for target location, communications, and intrusion alarm. It permits two-way conversations that cannot be monitored by an enemy.

10. An advance in laser technology makes possible three-dimensional pictures, or holograms, of large, motionless objects, for example, to record blueprints, engineering drawings, and other large documents for storage without first microfilming them; and to make optical computer memories with very high information capacities. Also foreseen are window-size holographic displays that would appear three-dimensional with field depths of at least 6 feet.

11. A high-speed printing process called Laserfax involves no impact such as pressing type or plates against paper. A laser beam strikes a hologram—a lensless photograph—and projects an image on a toner (powdered ink) paper. The thermal energy in the image fuses the toner to paper so that the image is printed. The first step in the method is to photograph the material on a transparency. The image is then transferred with the aid of the laser beam to the hologram, which serves as a permanent plate.

12. A laser beam used as a "track inspector" sweeps a railroad roadbed to detect foreign objects such as rocks, trees, etc., and a signal is flashed to a central station that warns approaching trains.

13. Holograms produced on a special magnetic surface through interaction of heat and light inherent in a laser beam can be erased magnetically. Scientists explain that the technique

could lead to development of an optical computer memory capable of storing 100 million bits of data in a film one-inch square that could be read out, erased, and reused repeatedly.

14. Selectavision, a color-TV player utilizing a very low-powered laser and holography for pre-recording programs on plastic tapes for reproduction through color TV sets.

SPACE OPTICS

Until a few years ago engineers considered it quite an accomplishment when the shortest wavelength radiations that could be produced coherently measured about a centimeter in length, while all other radiations at shorter wavelengths were incoherent as they originated from their source.

Development of the laser changed all this, observes Thomas P. Fahy, manager, technical operations, of the Electro-Optical Division of Perkin-Elmer Corporation.[7] The laser, he points out, permits extension to the light wavelength region of all the techniques that have proved to be of such economic and scientific importance in the radio and microwave fields.

"Lasers produce a beam of light so uniform in wavelength and polarization that they can be modulated to carry voice, music, and television transmissions," explained Mr. Fahy. "The intensity and coherence of laser light, whose frequency of radiation is 10,000 times higher than microwave radiation, appears to offer a means for sending message pulses superior to electric circuitry, radar, or microwave transmission.

"The ultraviolet and infrared bands of the light spectrum offer a vast new untapped medium not only for communications but also for power transmission. The intensity of the laser beams, coupled with their high directionality, appears to make it possible to send powers of millions of watts over non-spreading beams only a few centimeters in diameter. If laser radiation could be efficiently converted at both ends of such a beam, as someday it surely will, then such a beam would take on many of the characteristics of a high-power transmission line—a straight cable in space without poles or insulation."

[7] *Air Force and Space Digest,* May 1963.

The laser as an outstanding innovation of the Space Age has stimulated a revolution in the science and art of optics, heralded as a burgeoning industry. Innovations in cameras for tracking and observation as well as for photography and television; new photographic telescopes, radiometers, spectrometers, and photometers that track vapor clouds of rockets and monitor the effects of high-altitude nuclear explosions are all participating as new eyes on the cosmos.

Astronomers at the Boyden Observatory, operated by the Harvard College Observatory, and five European observatories in Bloemfontein, South Africa, succeeded in photographing the first Syncom satellite when it was "lost" at an altitude of 22,300 miles. Since the 86-pound craft was less than 3 feet in diameter, the observation by reflected light was considered a remarkable feat. Its brightness was only of the seventeenth magnitude, about 25,000 times fainter than the dimmest object that can be seen by the naked eye.

When Major Cooper's Faith 7 orbited over Africa, the Smithsonian Astrophysical Observatory reported that the spacecraft had been photographed by the Baker-Nunn tracking camera at the Smithsonian station near Johannesburg.

LASER PHOTOGRAPHY

Since that time the laser took on a radically new role in photographic optics, known as holography, which promises to extend the range of the astronomer's vision. But it does not stop there. Among the diverse possibilities is three-dimensional vision in numerous fields related to optics including television, motion pictures, radar, surveying, and microscopy for which a holographic microscope already has been developed by American Optical.

The first clue to this form of photography was discovered in 1801 by Thomas Young in England. He observed that if a pencil-like beam of light was split into two beams, and if those beams were brought back together on a screen by means of mirrors, a pattern of dark and bright bands appeared.

Seeking to increase magnification of the electron microscope in 1947, Dr. Dennis Gabor, of the Imperial College of Science and Technology in London, and later a staff scientist at CBS Laboratories, applied what is known as the "interference principle," which is the essence of holography. He is credited, therefore, as having discovered the art in rudimentary form. Efforts to advance the system were deterred by lack of adequate "tools," especially a source of coherent light, or light waves all in phase, which is essential in holography. Invention of the laser proved to be the key to practical advance, and holography was designated as one of the most potential offsprings of the laser.

A widespread resurgence of interest ensued in 1962 when Dr. Emmett N. Leith and Juris Upatnieks at the University of Michigan applied the laser as a source of coherent light and achieved amazing three-dimensional pictures called holograms from the Greek word "holos" meaning whole, inasmuch as the whole picture is recorded. Since then more than 100 laboratories have reported to be working on holographic projects and all sorts of applications. With numerous laser manufacturers already in the field it is recognized that holography brings to optics the dynamism of the electronics industry.

What is the technique? Holography instead of recording a visible image records reflected light waves. The hologram thus obtained contains no visible image. It is described as a hodgepodge of specks, blobs, and whorls, referred to as "a kind of optical code." It is a coded wave pattern produced from all the reflected light from an object. Leith and Upatnieks showed how a three-dimensional representation of an object is obtained by using a double laser beam to illuminate simultaneously the object and a light-sensitive photographic plate. The beam shining through the emulsified plate causes the latent image encoded in the emulsion to be reconstructed, or reconstituted, by a wave-front technique with remarkable three-dimensional realism.

The future of holography is foreseen as tremendous, unlimited, and portentous. It makes possible not only the creation of three-dimensional images but storage of visual data with a density never before achieved. It is estimated that the storage capacity is about 1,000 times greater than present magnetic memories, which means that the text of 10,000 average-size books may be stored in one cubic inch of film.

By holography, the laser makes it possible to store 1,000 standard-size book pages in a crystal 2 inches square and a quarter inch thick. Application for a patent on such a system was filed by Dr. Arthur N.

Carson, of Carson Laboratories, Bristol, Connecticut. Thus the headline LASER SHRINKS LIBRARY TO A CUBBYHOLE points to the future.

Endowed with fantastic image-data memory, holography challenges magnetic memories for a place in computers and mass-memory systems which appear to have what scientists foresee as "an infinite future of ebullient growth."

THE LASER FAMILY GROWS

Since success was achieved in making a ruby rod emit coherent light, scores of materials have been found to lase. An intense search is on for materials capable of that unique property. The hunt is reminiscent of Edison's long, world-wide pursuit of a suitable material for the filament of the electric lamp.

Helium, neon, argon, xenon, krypton, cesium, dysprosium, uranium, neodymium, gallium arsenide, and many other elements and materials are under test for ability to lase and extend development of the laser. Each laser material emits its light at one or more specific wavelengths. Gas lasers show considerable promise, since they can be designed to produce beams over a wide range of wavelengths. Various gases and gas mixtures have achieved maser action at approximately 150 different wavelengths, extending from inside the visible spectrum into infrared—and more are in prospect.

"New materials that lase are being discovered almost every week," said an engineer. "Radical advances literally pop into being every few days. It's really amazing. Now laser materials include crystals other than ruby—glasses, plastics, liquids, gases, and plasmas. Each must possess an atomic state with a suitable range of energy levels. For example, in the case of the gas laser, the active medium is a combination of helium and neon; operation depends on excitation of the neon atoms by collision with excited helium atoms. Based upon materials that lase, there must be a hundred different lasers and the number is certain to increase."

There is a glass laser—a high-quality optical glass infused with various fluorescent elements such as neodymium. Flame-type lasers may be possible, while scientists also are said to suspect that even a tube of air may be made to lase.

A liquid laser, using organic liquids and emitting coherent light at

13 wavelengths ranging from 7,430 to 9,630 angstroms, previously not available to laser action, has been developed at the Hughes Aircraft Company. These lasers, which are expected to be important to the fundamental understanding of matter, operate on a new principle—stimulated "Raman" scattering—observed in benzene and six other organic ring liquids.[8] In the ordinary Raman effect, well known to scientists, light is scattered from molecules. But this is said to be the first time the effect has been involved in laser action. The Raman effect is based on the fact that outgoing, or scattered, light has different energy and wavelengths than incoming light, the energy difference having been converted to molecular vibrations.

Previous lasers made use of fluorescence from a long-lived upper energy level of an atom or molecule. In the new laser materials, all of which are known to be strongly "Raman active," there is no upper energy level; and initial excitement of the material to an upper level is not required as in previous lasers. A very strong incident (or pump) light is required, however, to initiate laser action. As a consequence it has proved expedient to "pump" the liquid organic lasers with a high-power short-pulse ruby laser.

Adding a further new dimension to the technology of lasers, RCA fashioned one of plastic—a fine fiber of transparent material containing traces of europium, a rare earth in which the laser action is achieved.[9] Since it produces coherent pulses of intense crimson light at the highest visible frequency yet known to have been achieved by laser action, research men believe plastic lasers may be developed that emit coherent light over the visible spectrum from infrared through ultraviolet. It is reported to be easy to make and may be produced for a few dollars, whereas synthetic ruby lasers cost several hundred dollars.

Still another laser uses europium incorporated in a liquid organic

[8] C. V. Raman, an Indian physicist, discovered in 1927 that a photon of light (a positively charged particle) is absorbed by a molecule and re-emitted at a lower frequency. That interaction of light became known as the Raman effect. In 1962, Eric J. Woodbury and Won K. Ng of Hughes Aircraft Company discovered Raman laser action—an enhanced form of the Raman effect. Since then more than 100 coherent light sources have been produced by Raman laser action in various materials.

[9] The plastic acts as a holder for molecules known as "chelates," or molecular pincer-like claws, which completely enclose each atom of europium.

chelate instead of plastic. And there is one that uses carborundum (silicon-carbide), which has a number of advantages: operates at room temperatures, requires a low threshold of energy, and radiates well up into the visible spectrum.

Magnesium fluoride doped with nickel ions also proves to be a laser material, which in addition to emitting coherent infrared light, generates vibrations called photons in the crystal lattice. It is unique in that it lases at a wavelength determined partly by vibrations of the crystal lattice near the nickel ions and partly by the electronic state of the ions, whereas in other lasers the wavelength of the emitted light is determined by electronic transitions—the fall of electrons from a higher to a lower energy level.

So new is the laser that its possibilities for development and application appear unlimited. For instance, a new method that involves a magnetic technique for tuning, modulating, or pulsing light inside the laser crystal before it is emitted is seen by scientists as a key to practical laser-radar and communications with light that can be pulsed 100,000 times a second.

Scientists at the Ford Motor Company scientific laboratory have reported success in changing the characteristic red light produced by lasers into different colors or frequencies. This, they explain, opens the way for laser light beams to be controlled as are microwaves in communications and radar. When the red laser beam is passed through a liquid such as nitrogen, the frequency is lowered, and the beam then can be passed through a crystal which doubles the frequency. By changing the liquid, it becomes possible to produce almost any desired frequency.

Dr. Charles H. Townes reported at a meeting of the American Physical Society on April 29, 1965, that a laser beam of light had been used to produce sound waves more than a million times higher in pitch than those audible to the human ear. As a result, experiments were disclosing new and puzzling properties in liquids and solids. And the phenomenon can be used to modulate light passing through transparent material already set in oscillation by other light waves. The frequency of these laser sound waves, he explained, is so high that they travel only a fraction of an inch before being absorbed. And Dr. Townes added, "We are catching up with our dreams."

CHEMICAL AND LIQUID LASERS

Chemical lasers pumped into action by chemical reactions instead of by the application of an external source of electric power open a fertile area for research and development. Experimenters are intrigued by the possibilities in this laboratory domain in which electronists invite the chemists to join in pioneering.

The chemical laser is self-pumping. Energy is released by the making and breaking of chemical bonds instead of by an intense source of light or electron bombardment. Investigators explain that the most promising chemical reactions are exothermic; that is, they produce heat. The violent reaction between hydrogen and chlorine is reported to create interesting results, while iodine-atom lasers are said to show signs of promise for producing high-power levels. One photon entering the laser tube has been found to give rise to ten billion photons at the other end.[10]

It is all so new that scientists stress the fact that the chemical laser is in an early stage of development, and they prefer to evaluate its advantages as potentialities rather than facts. Also, they are aware of inherent limitations to be overcome. One challenge is to develop a chemical laser that will operate continuously. And there are many chemical reactions that call for investigation to determine their capabilities to lase.

While the search continues for gases and solids that will lase, some researchers follow another approach—liquid lasers, which are found to have a number of advantages minus deficiencies inherent in the solid and gas systems. First of all a solvent to qualify for laser use must show luminescence in the liquid state. Since a liquid laser was developed at General Telephone & Electronics in January 1963, various elements have been under study and advances are reported to presage competition with other forms of lasers.

SOLID-STATE LASERS

Scientists at the Lincoln Laboratory of the Massachusetts Institute of Technology discovered in 1962 that a simple junction diode

[10] Photon—a quantum of radiant energy moving with the velocity of light and an energy proportional to its frequency.

comprising a crystal of semiconducting gallium arsenide was an efficient emitter of infrared radiation. That focused attention on solid-state, light-emitting semiconductors, or electro-luminescence. Since then several laboratories have announced success in obtaining laser light from gallium arsenide diodes; other semiconductor compounds also are reported to qualify for service in lasers. In fact, as the transistor took over many duties of electron tubes so it is believed that light-emitting semiconductors might replace small incandescent lamps.

ORGANIC LASERS

Discovery of laser emission from complex organic-dye molecules was made in 1966 by physicists at the Thomas J. Watson Research Center. A feature is that a large number of such dyes, each fluorescing at a specific wavelength, are available over the range of the visible spectrum. This, the scientists explain, makes it possible to design a laser that emits coherent light at any visible wavelength. And it is pointed out that even more remarkable is the fact that each such laser is individually tunable; its output can be changed continuously over a smaller range of wavelengths; therefore the primary beam itself can be continuously tuned.

Widespread Research   Throughout the world the laser has fired the imagination of the scientific and engineering community. In fact, both scientists and engineers declare it is a real problem to decide which laser application to follow—there are so many.

Popularly defined as a magical optical-electronic device scintillating with countless possibilities, the laser is a beacon guiding half the laboratories of the world into the future. Seldom have the potentialities of an invention been grasped by so many so quickly. In the United States alone hundreds of research groups and industrial and university laboratories are engaged in laser research and engineering, as well as the Army, Navy, and Air Force. Equally important, thousands of youths interested in science have taken to study and research of the laser. Fascinated by its versatility and revolutionary possibilities, they foresee a future lased in new wonders in communications and in many other fields.

The carbon dioxide laser which produced an infrared beam with a power output of more than 8 kilowatts, the most powerful laser beam up to 1968, was credited with opening undreamed of areas for scientific investigation of great potential. It was considered to be particularly promising for radar and both terrestrial and extra-terrestrial communication since infrared beams are only slightly absorbed by the atmosphere.

The laser, however, is so new that it has had to make its own way in much the same manner as DeForest's audion, which represented a new technology the range and uses of which required development. Transistors, on the other hand, inherited a ready-made market. The test was whether a semiconductor could replace or supplement an electron tube to advantage—and in many applications it could. As a result semiconductors became a big segment of the electronic industry in a remarkably short time, and the growth continues. So too the laser is expected to advance with estimates already classifying it as a $1-billion-a-year industry in the "Space-Sailing Seventies."[11]

[11] Lasers were estimated to have become a $300-million-a-year business in 1968 and were foreseen growing to $1 billion annually by 1975 as the multiplicity of applications expanded.

XIV   New Ideas, New Inventions

> There is no law, no principle, based on
> past practice, which may not be over
> thrown in a moment by the arising of a
> new condition or the invention of a new
> material.
>
> —*John Ruskin*

SCIENCE ORBITING FAR INTO INTERSTELLAR REGIONS IS CULTIVATING
new ideas as it discovers the necessity for scientific advances vital for
successful technological conquest and leadership in the Space Age.

Shriveling time and shrinking distances, communications electronically makes the world a neighborhood through sound and sight while
the airplane mechanically links continents and nations, cities and
people. Now on an interplanetary scale the influence of distance is
being diminished across the universe just as radio leaped barriers of
the past to span the hemispheres.

An excellent illustration of how space stimulates thinking along
new lines is seen in a proposal by Dr. Freeman J. Dyson, at the
Institute for Advanced Study, Princeton, New Jersey, that gravity
machines using gravitational energy instead of sunlight as their main
source of power may someday be possible. It is pointed out that aside
from nuclear energy, the power that drives the world, whether coal,
oil, or water, originally came from sunlight. So gravitational energy
derived from the stars would be a radical innovation.

FORESHADOWING OF CHANGES

Despite the remarkable and revolutionary changes that David Sarnoff has witnessed in communications since his entry into the field in 1906, he declared in 1963, "We are just at the beginning of most promising developments."[1] And he added: "Our grandchildren's world will be one in which it will be possible to communicate with anyone, anywhere, at any time, by voice, sight, or written message, separately or as a combination of all three. Manned satellites weighing up to 150 tons and hovering over fixed points on earth will serve as switchboards in space to route telephone, radio, television, and other information, from country to country, continent to continent, and from the earth to space vehicles and planets. Participants will sit in their homes and offices, in full sight and hearing of each other through small desk instruments and a color TV screen on the wall.

"Within the next 10 to 20 years, it is more than probable that satellite television will be able to transmit on a world-wide basis, directly to the home without the need of intermediate ground stations. This holds enormous significance for people everywhere in entertainment, information, and education. Audiences of a billion people may be watching the same program at the same time while automatic language translators provide instant comprehension of the program's content.

" 'Ultimate' is a hazardous word to use in describing the future of any branch of science," continued Sarnoff. "If it has any application in the science of communications, it will probably arrive when an individual carrying a vest-pocket transmitter-receiver will connect by radio with a nearby switchboard and be able to see and speak via satellite with any similarly equipped individual anywhere on this or other planets.

"Ultra-high and microwave radio frequencies and laser beams can provide the billions of channels necessary for such personal communications. Private frequencies will then be assigned in much the same manner that an individual today receives his personal telephone

[1] Speaking before the American Friends of the Hebrew University, December 1, 1963, New York City.

number. . . . Not to recognize the basic forces of science and their impact upon society is to invite comparison with Aristotle's response when he was asked how much educated men are superior to the uneducated: 'As much as the living are to the dead.' "

NEW THEORIES, NEW PIONEERS

With the satellites come new theories, new systems, and new ideas that bring new men of science to the forefront. Among them, is James Alfred Van Allen, of the State University of Iowa, who, in experiments with rockets and satellites, in 1958, discovered clues that demonstrated the existence of belts of radiation that arched around the globe as a possible menace to spacemen; they were called the Van Allen belts, the inner one estimated to be 1,400 to 3,400 miles out in space and the other 8,000 to 12,000 miles.

Scientific data, however, transmitted by the Explorer 12 satellite, as analyzed and presented at NASA, indicated that the earth has but one huge radiation belt, not two, and that the unsymmetrical belt, earth-encircling to an elevation of 30,000 to 40,000 miles, may be blown by solar winds out as far as 100,000 miles or more from the earth. Scientists refer to this region as the magnetosphere—an extension of the earth's atmosphere, where atomic particles, because of their electrical charge, become trapped in the earth's magnetic field as iron filings cluster around a magnet.

SONG OF THE UNIVERSE

Harvard University's physics professor Edward M. Purcell was the first to detect that 21-centimeter waves (radio frequency of 1,420 megacycles) had their origin in cold hydrogen in space. The unwavering drone-like note of these 21-centimeter waves is said to re-echo throughout space as "the song of the universe" that comes from far beyond the range of the most powerful optical telescope. It is noticed that space is remarkably transparent to this frequency, making it especially useful in radio astronomy. And, since such waves are the sharpest and most universal in space, scientists suggest that they may prove to be ideal for interspacial signaling and interstellar exploration.

PLASMA PHYSICS

Greatly intensified is the study of *plasma physics,* which relates to ionized gases in which an important fraction of the molecules is dissociated into ions and electrons while the gas as a whole remains electrically neutral. The tremendous scope and significance of this vast field is evident in the recognition by scientists that the principal state of matter of the universe is neither solid, liquid, nor gaseous, but *plasma*—a system that includes many free electrons and ionized atoms the mutual reactions of which greatly affect its properties.

Astrophysicists refer to disturbances radiating from the sun, to the aurora borealis, and to lightning discharges as plasma phenomena. And the ionosphere—an ionized region created by solar ultraviolet light—is called a naturally occurring plasma, which has profound influence on communications through space.

Sir William Crookes, of cathode-ray fame, observed in 1879 that electrical discharges contained "matter in a fourth state or condition which is far removed from the state of a gas as a gas is from a liquid."

Before the idea could be probed and diagnosed, however, new technologies, efficient vacuum pumps, improved electronic research tools, and other scientific devices were needed. During the twenties Dr. Irving Langmuir and colleagues in General Electric's "House of Magic" developed methods to study the nature and action of particles in a mercury vapor discharge. In 1929, Langmuir and Dr. Lewis Tonks published a paper reporting on their observation that groups of electrons in a plasma could be made to oscillate in unison. The phenomenon was described as like the quivering of protoplasmic jelly—thus the name plasma.

They identified and described plasma as composed of equal numbers of negatively charged electrons and positively charged ions held together by their own electrical forces. They are generated when the atoms of a stable element are excited, generally by heat or electrical energy, to a point where their nuclei (ions) and some of their electrons dissociate and swarm in each others' vicinity. But since they are not energetic enough to escape their mutual electrical attraction, they form a seething, vibrating electronic force which is stable, cohesive, and electrically neutral.

Physicists have determined that plasmas vibrate at a constant frequency that varies with their density. And since vibrating electric charges can emit electromagnetic radiation ranging from radio waves to light, a collection of such charges in a plasma also can generate electromagnetic waves. For example, solid-state plasmas have the ability to generate high-frequency waves, and therefore it is foreseen that plasmas may lead to some radically new advances in communications.

Engineers explain that when plasma electrons are "kicked" from an exhaust, they drag ions along with them and a plasma jet is created for propulsion. Plasma engines, economical in weight and size, are not powerful enough in thrust to get a rocket off the pad, but once a chemical rocket puts the engine into space it can drive a space vehicle at terrific speed far out into the solar system—deep into the black vacuum of space.

Plasma propulsion of space vehicles equipped with electrical engines utilizing ionic or plasma acceleration appropriate for interplanetary travel is reported to be in the offing for orbital testing. Such ion engines are appraised by scientists as the ultimate in propulsion devices, which they believe eventually could send a manned spacecraft to Mars at a speed of perhaps millions of miles a day. The primary source of power is provided by a nuclear reactor, the energy of which is converted to electricity through a heat exchanger and electric generator. One useful propellant is cesium, the atoms of which are easily converted into ions.

Engineers assert that the concept of electrical propulsion is fast becoming an engineering reality. They foresee the application of the properties of plasmas, resulting in a profusion of practical uses including electrical, or magneto-hydrodynamic, generators with no mechanical parts, yet capable of producing vast amounts of power with nothing moving except electric charges.

Generation of electricity by blowing hot gas (plasma) between the poles of a powerful magnet fascinates and challenges physicists. Where conventional electrical generators use copper wire conductors rotated in a magnetic field to generate electricity, an MHD (magneto-hydrodynamic generator) utilizes a hot gas, ionized by having electrons knocked off some of its atoms to make the gas conductive. The term MPD (magnetoplasma dynamics) also is used. When the plasma shoots between the poles of a magnet at the terrific rate of 3,000 to 4,000 feet per second, a current flows across the flaming stream that may have a temperature of more than 5,000 degrees Fahrenheit; and the electricity so generated is picked up by electrodes.

Although these generators are still in the early stages of research, industrial laboratories are intensely conducting development work with a practical method as the goal. They are also vitally interested in liquid hydrogen engines, foreseen as another prospect as "new power for space."[2]

The goals of plasma physics research also include numerous new semiconductor devices, generation of electromagnetic waves, microwave amplification, and eventually an almost inexhaustible supply of electrical power using thermonuclear fusion reactions. "Plasma" becomes the password for a powerful thrust across new frontiers of physics that promises to put man far out into space and lead to innumerable new and revolutionary earth-bound developments in transportation as well as in communications. No wonder scientists affirm that plasma—the fourth state of matter—will be to the future what combustion has been to the past.

BIONICS

Bionics is another new branch of science based on development of artificial instruments that function like living organs or systems. For instance, a data processing device that duplicates the computer-like optical system of the frog has been built by RCA for experiments by the U.S. Air Force. Engineers believe that it may ultimately lead to a means of providing data interpretation—even decisions—in a variety of fields including air traffic control and missile detection.

Research revealed that the frog's bulging eyes are not only a visual

---

[2] First operation of a rocket engine propelled by electrical particles (ions) featured a one-hour trajectory flight of spacecraft Sert (Space Electric Rocket Test) conducted by NASA on July 20, 1964, from Wallops Island, Virginia.

or photographic system. It seems that the retina also acts as a natural analogue computer enabling the eyes to make quick decisions without bothering the brain. Anything unimportant is screened out; only those things that directly concern the frog are transmitted over nerve fibers to the brain. For example, an approaching insect or fly, which he may catch, is important to the frog as food. His brain never "sees" a fly moving out of range. A sudden shadow, however, may warn of a threat; if a bird swoops down, the retina flashes the alarm and the frog dives to safety.

As a research tool in the field of pattern recognition the electronic "frog's eye" is 6 feet long, weighs several hundred pounds, and has 33,000 components. Someday it may be miniaturized to detect missiles, make decisions on what action to take, intercept targets and order their destruction—all automatically in the twinkle of an eye.

The horizon of electronics does not end with this new leap of the frog into science any more than it did when twitching frogs' legs led Galvani to discover "animal electricity," and to the revelation of electricity as a current that flows.

Today, production of electrical power by direct conversion from light, heat, and chemical energy shows promise of becoming a flourishing domain for expansion of the electronics industry. Silently operating with no moving parts, mobile electronic packages are envisaged that will supply electricity to millions of people and to regions remote from power sources. Scientists believe it will be technically possible before the end of this century for electricity so created to serve any conceivable applications on earth, beneath the seas, and in outer space. Manifold benefits are expected to result for communications.

NEW PRODUCTS BORN OF SCIENCE

Inventions and discoveries that found no practical application when patented years ago are coming into their own because space exploration, communications, and modern industry need them. For instance, in 1911, when H. K. Onnes, a Dutch scientist, discovered superconductivity it was regarded as merely a scientific curiosity. He demonstrated the phenomenon by cooling a ring of lead or tin wire to nearly absolute zero. Then by inducing an electric current it would flow indefinitely without further injection of energy, because there was no resistance.

Conventional magnets require a steady flow of electricity, whereas magnets to which superconductivity is applied operate more economically; and for computers, superconducting promises major advances in speed, compactness, and economy of operation. The widespread possibilities extend to motors, the electron microscope, and many other devices used on earth and in space.

Electronic research men, engineers, physicists, and chemists are "on the go." At the Lincoln Laboratory of the Massachusetts Institute of Technology a diode has been developed to convert information-carrying electric current directly into light with almost 100 per cent efficiency. Using gallium arsenide as a semiconductor, instead of the germanium and silicon more commonly employed, this device has an active area about the size of a pinhead; in fact, it is described as a less-than-gnat-sized device. It generates pure infrared light when a direct current passes through it. And the intensity of the invisible infrared beam can be controlled by varying the strength of the current that generates it. In tests across 34 miles between Mount Wachusett and the laboratory at Lexington, Massachusetts, a Boston TV station was tuned in and the pictures were retransmitted on infrared rays. It is estimated that one circuit using the diode can carry 20 television channels or 20,000 telephone calls at one time. In effect it is based upon the sniperscope and snooperscope developed during World War II to see objects in the dark.

Space and military demands also have led to development of small, dependable power units—lightweight electrochemical batteries compared with the automobile's heavier lead-acid storage cells. And these new power units make possible cordless devices extending from radio transmitters in spacecraft to such consumer products as electric razors, vacuum cleaners, lamps, and a wide assortment of portable instruments including TV and radio sets.

HOMEWARD BOUND

Pointing out that electronics had been slow in going into the home, except in radio-television, General Electric announced in July 1965 that a Space Age component SCR (silicon-controlled rectifier), described as a cousin of the transistor, was ready for use in a range of products from automobiles to home appliances.

"The applications are limited only by the imagination," said Dr. L. C. Maier, Jr., "and extend to any instance where it is more convenient and desirable to select the exact level of speed, light, and heat in infinite degrees on your range—or any other consumer appliance—in the same way that you dial the exact degree of sound from radio or television."

ELECTRONIC TIMEKEEPERS

Should thousands of outmoded but garrulous solar-cell spacecraft get into space, they would clutter up the radio spectrum and tie up valuable wavelengths. Fortunately, it is not necessary that they "talk" all the time. To give their electronic "tongues" a rest, electronic timers have been developed that turn off a satellite's transmitter at any desired interval. Similar devices turn instruments on and off automatically so that satellites may take periodic readings of conditions in space over a long period without demanding exclusive and constant use of a radio channel.

INTEGRATED ELECTRONICS

In the late fifties, electronic and communications engineers had miniaturization as a goal. Demands of the Space Age in the sixties extended the objective to microminiature and even ultra-microminiature. A new technology—integrated electronics—evolved around thin-film circuits made by depositing or evaporating metals and other materials on glass or metal. As a result designers of computers and other electronic systems were enabled to work with components so tiny that millions of them could be aligned in a cubic foot. The gossamer-like circuits, along with miniature transistors hermetically sealed in tiny cans, led scientists to predict that the electronics industry is destined technologically and economically for revolutionary advances.

The fact that the microminiature circuits comprise passive elements such as resistors, capacitors, and insulators as well as active elements with the capacity of tubes, diodes, and transistors permit simultaneous, single-process fabrication of an entire thin-film circuit. It is an exotic field permeated by what is described as "a new kind of hysteria" that excites the imagination and stirs predictions that the most startling applications of this remarkable technology have not even been conjectured. Physicists declared that they are dealing in subdivisions of space never before encountered in industrial production; in fact, a million components can be packaged in a cubic foot, thereby reducing the size and cost of electronic instruments.

Indicative of the trend is a novel memory unit smaller than a pack of book matches that can process 100,000 computer words a second! And a miniature 12-pound magnetic tape recorder-reproducer that can record data continuously for 4 hours and play it back in 11 minutes has been developed for use in multi-orbit, manned spacecraft. It can record 74 million bits of telemetry data at 5,120 bits a second. The term "bit" is explained as follows: in the binary numbering system used in many computers only two marks (0 and 1) are used. Each of these marks is called a binary digit or bit.

Another innovation from RCA Laboratories is a solid-state element that combines the best properties of transistors and electron tubes plus unique features of its own. As a fundamental new building block it promises a wholly new kind of integrated microelectronic circuit. It permits up to 850 components to be arrayed in an area the size of a dime. Engineers and designers foresee a great future for this device, including portable, battery-operated, high-speed computers; lightweight, high-performance communications systems, and a new generation of tactical and industrial equipment operating over wide temperature ranges and with greater resistance to nuclear radiation.

A self-healing, radiation-resistant silicon solar cell is designed so that the spacecraft power source can operate indefinitely in radiation belts without damage. Application of the technique is foreseen for other devices used in space electronics such as diodes, transistors, and integrated circuits, making them immune to radiation.

"Videoscan" is the name of a new unit designed to examine 90,000 documents an hour and feed the data to a computer system for processing. This "optical character reader" handles a wide range of paperwork from gas, electric, and telephone bills to insurance premium statements, magazine subscription forms, tax notices, and many other similar items.

"Lasecon," a name derived from "laser" and "converter," is a new microwave phototube that combines high sensitivity with the extremely wide bandwidth of a traveling-wave tube. It is expected to accelerate the use of lasers in space communications, including optical radar, radar astronomy, and television.

Portending further glory for the electron tube, an experimental device has been developed that may open new channels of communication and enhance very high-definition radar. Operating near infrared light frequencies, this advance pushes up to the millimeter wave spectrum, well beyond the microwave ceiling. Signals oscillating 23 billion times a second are amplified. By comparison, TV signals oscillate from 500 to 800 million times a second.

Space graphology is a new technique made possible by RSA (radar signature analysis) which determines the shape of objects in space. Radar reflections trace the pattern. By measuring the strength of the reflected pulses the size and shape of the satellite, or missile, is revealed; also how fast it may be rolling, tumbling, or spinning. In combination with a computer the RSA analysis is speeded. Thus once in orbit a spacecraft is no secret as each vehicle traces its distinctive signature in the sky.

An electric power pack for spacecraft has been developed at RCA Laboratories based on thermocoupling. Two dissimilar metals of a thermocouple transform heat into electricity. A thermoelectric material (silicon-germanium) produces electricity at high temperatures. There are no moving parts and devices embodying the alloy are estimated to offer reliable electric power for at least five years. The heat may be derived from reactors, radioisotopes, the sun, or ordinary fuels. Laboratory tests are said to indicate that such a generator with a square-foot surface heated to 1,800 degrees Fahrenheit would produce nearly three times the electricity consumed at any one time in the average home.

A TUBELESS CAMERA

With an eye on the future when lunar astronauts will need a compact TV camera, RCA scientists have developed an experimental tubeless camera for which many applications loom both on earth and in space. Smaller than a man's hand it might be a key to an era of personal television communication. For the military, police, firemen, and newsmen, such a device could perform handy service. The applications in industry, education, and the medical field are manifold. The battery-operated transistorized unit features an array of about 132,000 thin-film elements deposited on four glass slides only an inch square to perform functions handled by standard TV pick-up tubes and circuits.

ALWAYS SOMETHING NEW

Already a multiplicity of new materials, instruments, systems, and even new industries adds up to show the tremendous thrust of the Space Age upon progress not only in electronics and communications but in related fields such as chemistry, physics, and metallurgy. New lubricants, new and tougher steels and superalloys that resist heat and corrosion are being produced.

Carbon fabrics developed for rocket nose cones and re-entry heat shields capable of withstanding temperatures as high as 5,800 degrees also have industrial applications for heat resistance, air filtration, insulation, electrostatic shielding, etc.

"Ancient" devices such as the conventional gyroscope with spinning wheels and ball bearings are being supplemented by gyros that have no wheels, bearings, rotors, springs, or bushings—no friction to cause drift from the proper direction. Among the innovations: cushions of hydrogen instead of ball bearings, and nuclear gyros that utilize atoms in which the nucleus behaves like a spinning top. In another form—electrically suspended—a rotor designed to spin for months in a rotating magnetic field overcomes friction and drift.

A laser gyro recognized as having ideal specifications for Space Age aeronautics is reported to show promise of replacing many conventional gyroscopes. It measures rotation in inertial space without use of a spinning mass as in conventional gyros. Therefore its performance is not affected by accelerations and it becomes a precision instrument of high accuracy. Its advantages include no special cooling requirement, low cost as well as low power consumption, and simplicity in construction. In design it is a simple laser arranged as a ring by using three or more reflectors. What of its future? "The indisputable choice over other more conventional sensors" is the engineering appraisal.

Thus the gyroscope invented and named in 1851 by Leon Foucault, French physicist, "to view the revolution of the earth" by making the rotation of the globe visible, is being revolutionized as "man's sixth sense" to detect his movements in space where there is no gravitation and therefore no sense of direction. The modern gyros, many in the processs of experimental development, attest to the validity of the remark that study and investigation lead scientists in the direction of discovery while practices and application carry engineers toward creation and commercial production.

COMPUTERIZED TYPESETTING

Electronic computers for automatic typesetting illustrate how the old order is changing. What served the past with distinction is being revitalized by invention and technological improvements to serve all forms of communication.

Mass media now are equipped with electronic "brains" and with an intricate nerve system of communications. News goes into space and to press in a symphony of sound—at least it's music to newsmen. Science has added many new instruments to augment the click of the telegraph and typewriter, the telephone voice, the buzz of wireless, the metallic clatter of linotypes, and the roar of the press, Electronics steps up the cadence. The radio-telephone, broadcasting, photoradio, television, teletypes, magnetic tapes, computers, and other innovations have turned the newspaper's city room into one that is international.

Technology accelerates news-gathering. The pace was slow when correspondents filed brief dispatches by telegraph or cable or mailed them.

Veteran journalists—Greeley, Bennett, Dana, Pulitzer, and Ochs —never heard of such things as the Digitronics-Dial-o-verter, a high-speed transmission machine that pours out copy, headlines, and messages at a rate that newsmen confess defies digestion. Coupled with electronics, teletypesetter tapes and photo machines have opened a new era of automation and speed in newspaper, magazine, and book publishing as well as commercial printing.

A unique electronic type composition system (RCA Videocomp, a metal-less typesetter) is capable of setting the entire text of a newspaper page in less than three minutes through the use of video and computer techniques together with magnetic tape. It "writes" 80 times more copy per inch of tape at speeds 300 times faster than can be achieved by punched paper tapes; and it is about 50 times faster than mechanical typesetters. A Colorscan system also has been developed to scan color transparencies and break them down into the four color separations required for full color reproduction.

Newspapers in various parts of the country are using an RCA computerized typesetting system. And IBM also has developed a computer system capable of fully automatic type composition. In one hour enough perforated tape is produced to set 12,000 lines of 8-point type. From the time the punched tape is created to the time hot type is poured by the automatic linecasting machine no manual operation is required.

In computerized typesetting, even hyphenated words at the end of a line are no problem, as every character and letter moves swiftly into proper place. The computer automatically incorporates all editing changes, counts each character, and keeps track of the number of spaces in each line so that the right-hand margin comes out even. It decides by logical analysis of vowels and consonants where split words are to be divided, and produces a second tape that is run through the linotype machine and which sets the type automatically. Thus the computer opens the way for greater centralization of newspaper production, and stirs predictions that electronics will revolutionize the composing room of the future.

Communication satellites prove the feasibility of world-wide automatic, high-speed, computerized typesetting, as demonstrated via the Relay satellite.[3] News copy from Chicago fed into an electronic computer at Camden, New Jersey, was transmitted through the satellite for automatic typesetting minutes later at the *Manchester Guardian, Glasgow Herald,* and *Edinburgh Scotsman.* Then the circuit was reversed, and news filed from London and Rio de Janeiro was relayed to the Camden computer, which translated the copy into taped instructions for transmission to Chicago for automatic type-

[3] Demonstration conducted on June 10, 1963, by RCA and Harris-Intertype Corporation in cooperation with NASA for the Production Management Conference of American Publishers' Association in Chicago, Illinois.

setting. From the time the copy left the point of origin until it appeared in print, it was controlled in every function by electronics.

SCIENCE EVER CHANGING

Science continually is providing new, more dependable and economical instruments and methods. Perhaps the communication satellites will confront competition. They are marvelous and expensive; it costs up to $13 million just to launch one. They have a relatively short operating life, although they may remain in orbit for a century or longer. No doubt, however, they will be simplified and made more foolproof to extend their electronic life span from months to years. And eventually spacecraft may be recoverable. The idea is to maneuver them back to earth for servicing and return them into orbit. In any event it seems reasonable to expect that international relaying will be revolutionized as it passes through the process of evolution impelled by research and engineering.

One ingenious possibility is to project a powerful laser beam high into space. Long columns of ionized or conducting air around the beam would act as a "sleeve" or "pipe" serving as a beamonized antenna. Affording great flexibility and ease of control, beams of this character might challenge elevated devices such as relay satellites.

LIQUID AND CULTURED CRYSTALS

Liquid crystals broke into the news in May 1968 when scientists at RCA Laboratories announced what they considered to be a research breakthrough of major importance for the future of electronic developments. The possibility of an entire range of new products is foreseen based on the control of light through liquid crystals. The broad scope of revolutionary applications for the future include: an all-electronic clock and wrist watches with no moving parts, highway warning signs, new radar screens, high-resolution electronic displays, even scoreboards, stock tickers, and eventually pocket-size TV receivers that could be viewed in bright sunlight. And the electronic soothsayers add, "Perhaps flat-screen mural television."

Indeed, a new technology was crystallizing that seemed destined to have a profound effect on electronics as a growth industry as old techniques were improved or altered, and no end of new ones introduced. No telling where the crystals might lead when linked with the laser and its coherent light beams!

Advancing on another front, RCA introduced a laminated construction technique to build transistors to rival electron tubes in power output, thereby opening the way for extended use of transistors in heavy-duty units such as all-solid-state sonar. Experimentally, success was announced in generation of radio waves oscillating at one million cycles per second with a power output of 800 watts.

By the time the transistor marked its twentieth anniversary in 1968, "crystal growing" had become a new art. Scientists were mastering the technique of creating single crystals of structural perfection and chemical purity surpassing materials produced in Nature which had been used in solid-state devices. Researchers pointed the way to development of cultured electronic components, including transistors, new forms of integrated circuits, and many other devices and systems. Success had been achieved in the laboratories in circumventing impurities, chemical interactions, and other factors that had plagued semiconductors. An intense effort got underway to reconstitute as many materials as possible into single crystal form that would lead to new and improved as well as more versatile components. In fact, it is declared that "increasing numbers of electronic components are no longer manufactured; they are cultivated like so many electronic pearls."

GLASSY ELECTRONICS

A radically different idea was presented by Stanford R. Ovshinsky, scientist-inventor of Troy, Michigan, at a news conference on November 10, 1968, several days prior to his publication on the subject in *Physical Review Letters,* journal of the American Physical Society. A front-page headline declared GLASSY ELECTRONIC DEVICE MAY SURPASS TRANSISTOR. The news featured the "Ovshinsky effect," discovered in the early 1960s, that puts simple, inexpensive amorphous glasses in the class of semiconductors. Physicists acclaimed the property as "the newest, biggest, and most exciting discovery in solid-state physics at the moment, representing totally new knowledge."

The inventor explained that electronic components smaller than one five-thousandth of an inch across, or one-third the size of the smallest transistor, could be made from the amorphous material. He foresaw many applications for "ovonic devices" including flat-screen television hung on the wall like a picture, desk-top computers, new switching devices, and missile guidance systems impervious to destruction by radiation.

XV   Advances in the Spectrum

> The voice of Time cries to man, Advance!
> —*Dickens*

ENGINEERS KNOW THAT RADIO WAVES, INFRARED AS WELL AS VISIBLE light, X-rays, and cosmic rays are all part of the spectrum of electromagnetic radiation. They say that any electromagnetic wave is a traveling disturbance in space or matter—an interplay between electric and magnetic fields.

The spectrum ranges in theory from zero frequency (infinitely long wavelength) to infinite frequency (zero wavelength). In communications the lower limit of frequency is about 10 kilocycles (long waves used for wireless), and the upper limit, about 56,000 megacycles (5 millimeter radar). Then comes a "dead band" between the shortest microwaves and infrared, with wavelengths in the band ranging from a few millimeters to about 30 microns (a micron is one-millionth of a meter).[1]

Since the advent of wireless, radio engineers have been confronted with various "dead bands" until eventually they developed techniques and instruments to make them useful for communications. Mindful of lessons taught by the past they are confident that 1-millimeter waves will become practical and important in communications. By pro-

---

[1] During 1966 the move to use the word "hertz" to supplant cycle as the standard unit of frequency gained momentum. Thus kilohertz (kHz) instead of kilocycle and megahertz (mHz) in place of megacycle would honor Heinrich Rudolph Hertz, discoverer of electromagnetic waves.

ducing coherent radiation—a train of waves under control—in the middle of the dead band it may no longer be dead. Hope along this line is derived from the prospect that an offspring of the maser or laser may penetrate the band and harness its potential advantages for communications on earth and in space.

THE IONOSPHERE

Until the advent of rockets, satellites, and orbiting laboratories, radio techniques applied from the ground had to be depended upon to glean knowledge of the ionosphere, a vast region of ionized gases. Technically it is defined as "an outer belt of the earth's atmosphere in which radiations from the sun ionize, or excite electrically, the atoms and molecules of the atmospheric gases." In effect it is the "roof" of the earth's atmosphere from which radio waves are reflected, causing them to hop, skip, and jump around the world as a boon to long-distance communication. Radio experiments that began in the twenties located the ionosphere from 30 to 250 miles up, but instrumented satellites indicate that it may extend up several thousand miles. Space Age research makes that region no longer an exclusive province of radio engineers; it becomes "an inseparable part of the whole mechanism of astrophysics."

Scientists attending the 1962 International Conference on the Ionosphere in London stressed the fact that satellites have shed new light on the chemistry of the atmosphere and on the mechanisms whereby the sun's rays and gases of the atmosphere interact to create various atmospheric layers.

Present thinking divides the ionosphere into four distinct layers, each of which reflects different wavelengths, depending upon the layer's electron density. Layer D exists only in daylight and ranges from 30 to 60 miles up. Layer C, described as "transient," is about 10 miles below D. Layer E is 55 to 90 miles high; and F, up 85 to 600 miles, is the main reflector of shortwave radio.

Without electrons in these regions, especially in E and F, scientists explain that it would not be possible to reflect radio waves from these "mirrors" in the sky to places beyond the horizon. By studying the reflections over a period of many years physicists have been able to estimate the density of electrons in each of the upper regions. Since electrons arise from the ionization of atoms and molecules, it is calculated that there must be an equivalent density of positively charged ions. But what kind of positive ions? Scientists are hopeful that satellites will send back the answer, for already the chemical structure of the atmosphere and ionosphere that emerged mainly from theory has been substantiated to a large degree by scientific instruments in orbit.

MICROWAVES

All regions of the radio spectrum have come under cultivation to meet the increased demands in communications both national and international. Large-scale adoption of microwaves by the common carriers accelerated widespread use of radio for domestic communications. In foreign lands, too, microwave systems have made impressive advances, overcoming such barriers to communications as mountains, oceans, jungles, Arctic, and Antarctic, as well as the weather.

When the FCC in 1960 made frequencies available to virtually any business or industry, a new market potential was opened for microwaves. Since then webs of communications have been woven over municipalities for police, fire, public safety, and highway departments, while the common carriers continually are extending microwave facilities. For example, A.T.&T. operates an 850,000-mile network of microwave radio relays that includes a multiplicity of telephone circuits. Western Union has a 7,500-mile microwave network that zigzags across the country, including various spurs that add 80 million new telegraph channel miles to its nationwide facilities. A leased "super-group" bandwidth expands the network another 5,300 miles. Coast-to-coast message traffic can be handled at the rate of 2.4 million words per minute, and in addition other services such as teleprinter, high-speed data, and facsimile.

Thousands upon thousands of miles of microwave networks are in use not only by the common carriers but by government agencies, the armed forces, and the radio-television industries. Utilities, airways, pipelines, and similar right-of-way businesses use microwaves, while taxis and delivery and service trucks use ultra-high frequency mobile radiophone for dispatching and to speed operations. And whether building skyscrapers, bridges, dams, or airports, construction crews find radio convenient for directing and coordinating the work.

Along the turnpikes and superhighways microwaves are called "guardian and expediter of the highway traveler," for they provide additional facilities for police, administrative, and maintenance communication as well as remote control of warning signs.

On railroads microwaves are a weatherproof, streamlined means of transmitting signals that control and activate signals, govern electric power flow, and perform many other chores such as speeding waybills between terminals by facsimile reproduction.

New link after link is being added to microwave systems throughout the world. Continued growth is foreseen in both commercial and military fields, with microwaves being brought into use for transmitting business data from scattered locations to central computing headquarters.

The optical qualities of these waves make it possible to focus them into narrow, powerful beams in a line of sight over long distances. Highly directional antennas focus the waves into a concentrated beam aimed at a fixed repeater station located on an average twenty-five miles away. From there the signal is relayed to the next station. Through the process of multiplexing, the microwaves derive tremendous capacity. Subdivided into comparatively narrow bands of frequencies the microwaves channel can handle simultaneously many types of information. And since the microwave is free of physical limitations of poles and wires, initial investment and maintenance costs generally are lower than for wire lines of comparable facilities. Immune to severe weather and air turbulence, microwave systems are likely to stretch beyond the atmosphere as the Space Age enlists their service.

A "quite unexpected" discovery that a tiny crystal of gallium arsenide emits microwaves when a steady voltage is applied across it was made by J. B. Gunn at the IBM Research Center in 1963. Requiring no vacuum or filament and less power to operate, it is much simpler, more compact, and less expensive than the klystron, which has dominated the microwave field as the electron tube did radio prior to invention of the transistor and other solid-state devices. Several other types of solid-state microwave generators which have appeared in addition to the Gunn oscillator lead engineers to foresee that a revolution in microwave technology may be in the offing opening the way for greater application and more widespread use of the tiny waves.

CITIZENS RADIO

Prior to the FCC's establishment, in 1960, of a "citizens band" of radio frequencies for use by the general public, the farmer, boat owner, and a host of others lacked the benefits of mobile communications. Two-way radio provides a new party line from farmhouse to tractor, from home and boatyard to boat. One has but to eavesdrop to realize that "the voice of the people is on the air." Proof of the popularity of the service is found in the fact that citizens radio installations exceed amateur radio transmitters in the United States.[2]

TROPOSCATTER

It was long believed that radio waves used for beaming messages and telephone calls from point to point were effective only for line-of-sight communications; they appeared to fade rapidly beyond the horizon. During World War II, however, it was noted that some radars picked up targets at distances theoretically well beyond the horizon, while microwave radio signals also turned up beyond the expected range. Studies of the phenomenon revealed that transmission beyond the horizon was possible and practical. As a result, a new high-quality communications service was developed for areas where conditions of terrain and weather prevented the establishment of frequent relay points.

Alaska, possessed of various factors that plague communications in the far north, was the first to benefit. In 1955, the Western Electric Company, under contract with the Air Materiel Command, began work for the U.S. Air Force on a radio relay network named "White Alice." In March 1958, it was placed in operation with 33 stations in the network covering 3,000 route miles, providing 170,000 telephone circuit miles and 50,000 telegraph circuit miles. Circuits are made available to the public through the Alaska Communication System, which provides civilian long-distance communications.

The clue to the problem of "beyond the horizon" communication that made "White Alice" possible was found in the troposphere, the lower layer of the earth's atmosphere, about 5 miles thick. It extends

[2] Citizens stations in 1969 totaled 875,125 and licensed amateur stations 275,000.

up about 60,000 feet at the equator and 30,000 feet at the poles. The technique is known as "troposcatter"—a diffused reflection or scattering in all directions of a radio beam aimed obliquely into the troposphere.[3]

Giant transmitting and receiving dish-shaped and "scoop" antennas, 60 feet tall and resembling outdoor movie screens, generally are perched atop mountains. Each antenna is aimed precisely toward the antennas of the next station, which may be as much as 170 miles away. When powerful radio signals leap from the transmitting antenna they shoot toward the horizon in a straight line. Traveling through the air, a tiny fraction of the radio wave is scattered downward in the troposphere. So faint is the received signal that it is rated only about one-ten-trillionth of the strength of the signal transmitted. It is amplified and relayed along the route by repeating the process of transmission, scatter, and pick up.

Professionally, the technique is called "forward propagation tropospheric scatter." It is being applied in various regions of the world, providing commercial and military communication facilities where they could not exist before, chiefly because of wild terrain and fierce weather. It is a feature in the Dew Line (Distant Early Warning) across the Arctic from Alaska to Greenland and on to England, as well as in the BMEWS (Ballistic Missile Early Warning System) network across the Arctic and the Atlantic. Several islands in the Pacific are linked by troposcatter, and a tropo link also has been used to carry TV programs between Miami and Havana.

RACEP, RADA, AND MADA

Challenged by the fact that the radio spectrum cannot be expanded, research engineers of the Martin Marietta Corporation concentrated on design of apparatus for its more efficient use. As a result they developed RACEP (Random Access and Correlation for Ex-

[3] Ionospheric scatter utilizes very high frequency (VHF) waves in a similar way except that the waves are scattered from the ionosphere about 50 miles above the earth, permitting transmission up to 1,200 miles. Tropospheric scatter using ultra-high frequencies (UHF) or superhigh frequencies (SHF) provides reliable, high-grade multi-channel communication between two points on the earth's surface from 70 to 600 miles apart. Tropo system projects are being developed throughout the world and are foreseen as complementary to satellite systems.

tended Performance), heralded as a breakthrough in advancing communications reliability and operational flexibility.

From RACEP evolved RADA (Random Access Discrete Address) and MADA (Multiple Access Discrete Address), the latter providing automatic access to all system channels, not just one fixed channel. That improves ability to communicate when needed without waiting. And in military tactical communications situations, elimination of the fixed frequency reduces the effects of interference, eavesdropping, or jamming.

The RADA system with a capacity for more than 40,000 discrete addresses—individual numbers assigned to a specific user or group—automatically selects the channel best adapted for each communication situation. Direct user-to-user communication is provided, incorporating the inherent mobility advantages of radio with the signaling capabilities of an automatic telephone system to handle voice, facsimile, teletypewriter, and data information. The system also provides conference calls, a priority override permitting certain commanders to pre-empt calls in progress, in addition to other features.

Army commanders are provided direct-dialing telephone-type communication with other divisions employing the same system and with all echelons of their command. No wires or bulky switching facilities limit mobility or deployment pattern of a division as in the past when fixed communications centers were deployed.

Among the military and civilian fields in which MADA techniques can serve are: Pilot-to-ground-controller communications in both civilian and military air traffic control. Voice and data may be handled simultaneously by the same equipment.

Communications for commercial firms and municipal departments (fire, police) between controller and dispatcher as well as fleet vehicles such as taxis and delivery trucks.

Communications in time of emergency disaster.

Private, commercial, and military ship-to-shore communications.

Private telephone-type world-wide communications via orbital satellite relay.

AUTODIN

The Space Age makes it imperative that the military, with its forces scattered around the world, have superior communications of all kinds at all times. To achieve this, Western Union, working with the Department of Defense and the Air Force, developed Autodin— the Automatic Digital Network of the Department of Defense—appraised as an advanced breakthrough in data communications.

The computer-controlled network basically consists of nine switching centers in the United States interconnecting more than 2,000 bases, air stations, depots, and other authorized installations in a global system for world-wide, high-speed interchange of vital data. Autodin handles more than 250 million messages annually—three times the number of messages carried by the public telegram system.

"Accuracy is so great that no more than one error in ten million characters can pass undetected," explained an operating engineer. "In the past, teletype machines, data card transmitters, and magnetic tape devices could transmit only to similar machines; but Autodin gives us a common electronic language so that all of these communication media can be used simultaneously within the same network. For example, it is possible to exchange data freely between punched paper tape, punched cards, and magnetic tapes from a variety of computers. All information transmitted is automatically encrypted and decoded at high speed at the destination.

"Autodin would have been impossible just a few years ago," he said, "since the first large-scale general purpose computer was developed in 1944, and the first electronic computer was not unveiled until 1946. Now computers operate with extraordinary reliability, with as many as a billion or ten billion operations without error, and speeds have gone up to more than 100,000 additions a second. . . . It is estimated that by 1970, at least one-half of all communications will be in data form, and that somewhat later communications between machines will exceed that between people. We are facing a revolution in automatic communications destined to greatly increase man's capabilities in planning, analyzing, computing, and controlling. Autodin is just the beginning—a new beginning in communication."[4]

Messages and data transmitted over the Autodin network have doubled in volume every two years since first placed in operation; new, faster, more versatile "third generation" computers in the

[4] The General Services Administration operates a network of 1,800 stations coast to coast, built by Western Union for the Advanced Record System of the Federal Telecommunications System serving civilian agencies of the Government.

switching centers will handle increased traffic in the seventies. Communications engineers in this Electronic Age are participating in what they call "the information revolution," and they point out that this is a moment in history when it becomes clear that a computer can do in minutes what communicators once did in weeks, months, and even years.

It becomes more and more apparent that, just as television engineers adopted and adapted everything they could from radio, so the astro-electronic engineers will continue to delve into every phase of electronic communications and harness whatever is available for interspatial signaling. They have found several important features in radio telemetry[5] and with minor modifications have applied those proven techniques to satellite communications. For instance, the huge parabolic dish antennas for tracking satellites and space vehicles as well as for reception of TV pictures and scientific data relayed by radio are products of such space services as radar, telemetry, and radio astronomy.

[5] Telemetering is the technique of recording data associated with some distant event, usually by an instrument reading being radioed from a plane, rocket, or satellite to a recording machine on the ground.

XVI   Exploring the Universe

> Ah, but a man's reach should exceed his
> grasp
> Or what's a heaven for?
>
> *—Browning*

Scientists exploring the universe from the surface of the earth have few "windows" through which they can "see," or receive messages from beyond the atmosphere, which is transparent only in limited regions of the electromagnetic spectrum. Through these "windows" comes visible light that enables them to see the cosmos, and also the invisible waves which, converted into sounds heard on radio, radar, and television, yield valuable scientific information concerning the planets and stars.

Since there are so many unsolved mysteries and so much activity in interplanetary and interstellar space that cannot be seen through these atmospheric windows, modern Galileos have had to develop other scientific means of collecting information which the atmosphere blocks from reaching the surface of the earth. They must penetrate radio-electronically the earth's shielding atmosphere, launch robot explorers, or go themselves to the moon, Mars, and other planets.

Solar telescopes and cameras already have been rocketed above the clouds and beyond the earth's atmosphere to glimpse and to photograph an eclipse of the sun. Stabilized in space long enough to make exposures, the cameras parachute to earth with their films.

Cosmic geography and astrophysics become new and fascinating subjects. Where schoolboys have studied the continents and oceans, rivers and mountains of the earth, the textbooks of the future are likely to cover radiation lagoons, magnetic reefs, atmospheric tides, the hydrogen sea, the cosmic ocean, and streams of rays that flow in space.

Cosmic particles are described as bits of matter—atomic nuclei—that constantly shoot in all directions through space and, it is estimated, through the body of every person on earth at the rate of some 30 every second without harmful effect. Eruptions of the sun have been known to release such particles; however, most appear to originate in star "explosions" known as supernovae, or else from sun-like stars. Scintillation counters detect them; and at times intense cosmic ray "showers" strike the atmosphere, indicative of huge cosmic ray storms in outer space.

While studying sources of static that interfere with radio reception, Karl G. Jansky, an American radio engineer, recognized in 1931 that one source of periodic "noise" came from electromagnetic waves originating in outer space. That discovery led the way to development of radio telescopes, giving man a new "window" through which to view the cosmos.

Twenty years later, in 1951, Walter Baade, using the 200-inch optical telescope on Mount Palomar, took pictures of Cygnus A (discovered in 1948) that identified it with a remote galaxy and revealed it to be an intense source of cosmic radio emissions. The pictures showed that Cygnus A coincided with the position of a visible galaxy, or possibly two galaxies in collision, estimated to be 700 million light years away.

In the boundless quest for knowledge of the cosmos, a new kind of astronomy developed. All over the world astronomers are scanning the heavens with radio-radar telescopes which facilitate the study of celestial objects through their natural electromagnetic radiations. These telescopes feature big dish-shaped antennas or reflectors. Their large dimension is generally necessary because the length of radio waves is millions of times that of light waves detected by optical telescopes. Therefore, the longer the wave under study, the larger the dish antenna needed.

While optical astronomers see, the radio-radar astronomers hear and, in a sense, touch objects in the heavens. They have charted a precise outline of the Milky Way; they have determined how the galaxy rotates; and they have discovered it is expanding.

Scientists at the Jodrell Bank Radio Astronomy Observatory, 25 miles south of Manchester, England, have succeeded in measuring part of the magnetic field of the Milky Way; and it is thought to be the first positive evidence of the existence of such a field. Astronomers recognize it as the force that holds the hydrogen out of which it is believed new stars are constantly being created.

WATCHING THE STARS AND PLANETS

The stars and the vast interstellar clouds of hydrogen are of course under close observation. All the planets are under radio-radar scrutiny. Indeed, one might well wonder if some of the strange noises coming from outer space are radio probes from some other civilization seeking to learn about the earth—its surface temperatures, life, and vegetation. Scientists at the University of Michigan, using a radio telescope and a ruby maser amplifier, have recorded the surface temperature of Saturn to be minus 280 degrees Fahrenheit.

There are two main sources of cosmological, static-like radio waves emitted from the planets: thermal radiation, emitted when a planet is exposed to intense heat; and non-thermal radiation, caused by disturbances such as flares in the region of the planet. Neptune, Uranus, and Pluto are yet to be heard. Pluto, however, is so cold and such a great distance away that some doubt exists that "signals" will be detected from it.

Radar provides radio astronomers with still another new tool. By reflecting signals off celestial bodies and by measuring the time required for the echo to return, observers believe they will be able to determine with fair accuracy what the terrain is like on the planets.

REVELATIONS FROM VENUS

Radar signals were bounced off Venus, in March 1961, by the Goldstone Tracking Station in California's Mojave Desert. The radar beam was aimed at the planet 35 million miles away by an 85-foot dish antenna, and 6.5 minutes later the echo came back. Objectives of this and other experiments follow:

To determine whether Venus spins on an axis, as the earth does, and if so, the speed of rotation, which would be revealed in a change in the frequency of the radar signals.

To determine the orientation of the planet's spin axis.

To investigate the nature of the planet's surface by comparing the radar echoes with those given off by certain materials on earth, such as water and rock.

To make more exact definition of the length of the astronomical unit, the basic measuring stick used by astronomers for calculating interplanetary distances. The astronomical unit, the mean distance from the earth to the sun, is about 93 million miles.

Does Venus have a day-night sequence? If not, one side would be torrid, the other frigid.

From an analysis of radar measurements and tentative discoveries, one group of observers reports evidence that Venus rotates slowly, probably once in its trip around the sun, which would be once every 225 earth-days, the length of a Venusian day. These results are not conclusive; another group interprets the radar signals as indicating Venus has little or no rotation. If that be true, one side of the planet would be continuously roasting in scalding sunlight and the other side frigid forever.

Astronomers at the Harvard College Observatory report having obtained a Venus temperature of 600 degrees Fahrenheit as revealed by radio emissions on the 21-centimeter wavelength. This would seem to substantiate evidence that at least one side of Venus is a torrid desert, certainly no place for astronautical explorers or space-craft. Other radio emissions detected from Venus are interpreted as indicating a temperature of 800 degrees Fahrenheit. Therefore, it is deduced that whatever the nature of the surface, it could not be wet.

Powerful radar beams aimed like a giant searchlight at Venus continue to penetrate the perpetual cloud cover and by reflection help to map the surface. Such rays mirrored off the Venusian terrain when the planet is within 30 million miles of the earth are analyzed by computers to reveal the topography. Through this technique two major features of the planet are reported to have been discovered, both believed to be mountain ranges—Alpha and Beta—comparable in; extent to the Rockies. The height, however, remains a mystery which radar may yet penetrate.

HAYSTACK

"Haystack" may appear to be an odd name for a modern complex antenna system that can probe 100 million miles above and beyond the Massachusetts farmlands. While scientists call it haystack, farmers in the neighborhood refer to it as "the big ball up on the hill" in which they understand "something is going on."

It comprises a multi-purpose experimental radio-radar facility for research over a broad range of space communications. Situated at the Millstone Hill Field Station near Tyngsboro about 30 miles northwest of Boston a mammoth metal-frame rigid radome 150 feet in diameter shelters a 120-foot-diameter antenna system reputed to be the most precise structure of its size ever built. It is a 680-ton instrument designed as an evolutionary research tool destined to constantly grow and change, never to be finished or rated as complete. Science moves too swiftly for that. The antenna beam is pointed automatically by a digital computer as instructed by an operator at a conventional typewriter keyboard which literally puts control of the huge, finely tuned antenna system at his fingertips.

Haystack's four major functions are: (1) To serve as a space-communication ground terminal. (2) To track and measure by radar. (3) To act as a radio telescope. (4) To aid in evaluation and development of advanced radio and radar techniques.

Some of the typical experimental uses include:

1. Relay communications via the moon and via passive and active satellites. Study the effects of the atmosphere and weather at extremely high frequencies. Transmit to and receive from space probes up to 100 million miles.

2. Precision radar tracking of small targets at great distances, illustrated by the fact that a target the size of a .22-caliber bullet could be tracked 1,000 miles, or a moderate-size satellite at ranges of more than 20,000 miles. The project includes detailed radar measurements of the moon, Venus, and other planets in order to study orbit perturbations and surface features.

3. In the field of radio-radar astronomy the very narrow beam and precise tracking capability provide the system with unprecedented sensitivity to study the atmospheres and surface properties of the moon, Venus, Jupiter, Mars, and Mercury. It also qualifies to map interstellar clouds of hydrogen and possibly locate and identify materials not previously detected. Still another mission is to prepare a comprehensive radio map of the center of the earth's galaxy.

NEW CONCEPTS OF SPACE

Scientists confess that within recent years they have changed their concept of outer space. It was believed that magnetic fields and other zones existed around the earth, and that a vast void extended between the earth and sun. Now the radio astronomer says, "The void is getting smaller as we poke around there." It is thought that the earth is located in the outer boundaries of the sun's various influences—with no real void between.

Scientists are hopeful that radio astronomy may offer evidence to support the steady-state theory of the origin of the universe, or the theory that the universe began with a big bang billions of years ago.

Both of these cosmological theories, however, are said to leave a big question unanswered. In the case of the big bang theory, the question is where did the material come from that came into existence at the time of the big bang, possibly 20 billion years ago? In the case of the steady-state theory, the question is where does the material that must go into constant creation come from?

Will radio-radar astronomy find the answer?

The idea of a universe born "from nothing" in a single explosion raises philosophical as well as scientific problems. Is the universe infinite or limited in extent? Is it eternal and unchanging? Was it created in a "big bang" or is it oscillating? Dr. Allan R. Sandage, through observations of quasars at Mt. Palomar observatory, believes he has seen evidence that the universe may be oscillating at a rate of one "bang" every 82 billion years.

Bell Laboratories' scientists, Dr. Arno A. Penzias and Dr. Robert W. Wilson, as well as a group at Princeton University led by Dr.

Robert H. Dickie, Professor of Physics, believe they have observed through radio-astronomy "what may be remnants" of a primordial explosion that gave birth to the universe. Such signals seem to imply support for the "big bang" theory and therefore create doubt regarding the steady-state theory. In any event, the philosophical as well as scientific riddles are likely to puzzle scientists from generation to generation.

QUASARS

Astronomers are intrigued by the identification of far-distant objects that look, superficially, like stars but actually appear to be enormously larger and much more brilliant. They emit light and radio energy with such great intensity that astronomers believe they are illuminated by physical processes unfamiliar to science. When discovered in 1963 they were called "quasi-stellar radio sources," but that nomenclature was abbreviated to "quasar."

Astronomer Maarten Schmidt of Caltech discovered a quasar calculated to be rushing away from the earth at 149,000 miles a second, 80 per cent the speed of light. It is said to be the most distant object ever identified. Astronomers say that quasars are by far the most brilliant objects in the universe, shining with the light of 50 to 100 galaxies, each containing 100 billion stars as bright as the sun. Where, ask the astronomers, did all the energy originate? And quasar light had to travel billions of light years before reaching the earth.

Scientists estimate that the quasars existed briefly and violently even before the sun was born, possibly after the birth of the expanding universe 15 billion years ago. Most of the 60 or more identified quasars are calculated to lie beyond all other visible objects, thus radio-astronomy must be depended upon to glean more knowledge concerning them.

It has been suggested that blasts of energy from quasar CTA-102 are coded messages from a super-civilization. Scientists in most instances, however, reject that theory on the basis that no civilization no matter how far advanced would be able to switch on and off a fantastic energy output of 10,000 billion suns. And they say that in the far-distant realm of the quasars, space itself may have unfamiliar properties. Their hope to know what quasars really are rests in the study of spectrograms that may reveal new knowledge, at least up to the edge of the universe, if such a frontier exists.

PULSARS

Pulsars, celestial sources of radio emissions, create another mystery that intrigues and puzzles astronomers. When the first four were discovered by radio astronomy—not optically—in 1967, the machine-like regularity of intense bursts of radio energy caused earthlings to wonder if another world might be signaling. A year later, radio telescopes at the National Radio Astronomy Observatory in West Virginia, and the Arecibo Ionospheric Observatory in Puerto Rico, detected a fifth pulsar throbbing 30 times a second in the Crab nebula some 6,000 light-years distant from the earth.

Preliminary observations lead astrophysicists to suspect that pulsars are very small, extremely dense objects rotating like white dwarfs, or neutron stars. And scientists theorize that a hot gas or plasma trapped by a neutron star's powerful magnetic field can generate electromagnetic energy. Cosmologists conjecture that both quasars and pulsars manifest a phenomenon known as "gravitational collapse" and the waves or signals may be gravitational radiation of an infinite range. To pursue investigations, experiments are being conducted with radically new forms of antennas designed to tune-in waves of gravitational radiation.

THE "SONG" OF HYDROGEN

How does hydrogen, the most universal and lightest of the basic elements, gain such predominance throughout the universe? Was it created by a tremendous bang and dispersed throughout space, or is it continuously being manufactured in the stars? These questions have been paramount since Henry Cavendish, in 1766, discovered what he termed "inflammable air." Antoine Lavoisier, distinguished chemist of the eighteenth century, identified it and coined the word hydrogen; he laid the foundation of modern chemical nomenclature and listed 33 chemical elements.

Now that scientists have learned to recognize "the song of the universe"—hydrogen's radiations as heard on radio—the problem is to interpret its meaning and to learn more about the structure of the universe, of which hydrogen is a basic building block.

The enormous power of hydrogen has greatly intensified efforts to harness its energy for work on a widespread scale encompassing many fields of life and industry. Long a key factor in chemistry and the chemical industry, its power and potential extend into the gas and oil industries, for hydrogen is fuel. Scientists agree that nothing equals liquid hydrogen for propulsion into space. Acclaimed as "the ultimate fuel," hydrogen, with its high combustion energy, gives a terrific thrust to rockets. It is expected to make possible hydrogen-powered planes for space flight, as well as nuclear rockets—advanced engines for space. By chemical and thermonuclear means, scientists believe that hydrogen, within the next 20 years, is destined to yield electrical power from almost boundless sources.

Although the simplest of the elements, hydrogen is appraised as a master key to future scientific and industrial advances on earth as well as in space. Fortunately, it is abundantly present; about 93 per cent of all matter by volume, and 76 per cent by weight, is hydrogen. It is an integral part of the universe and is credited with supplying almost all the energy generated by the sun and stars.

Physicists describe the hydrogen atom as composed of a single proton (a positively charged particle) with one electron revolving about it as a satellite whirls around the earth. In the thirties, Linus Pauling, of the California Institute of Technology, discovered the electronic nature of the chemical bond, beginning with hydrogen, for which he received the Nobel Prize in 1954. It was observed that the hydrogen molecule is bound in a bond formed by the sharing of the two electrons between the two hydrogen atoms, and this is said to be one of the most important unions in nature. Scientists have found it easy to break the bond and thus provide energy in reaction, and they find it easy to open the coupling to effect a combination with other elements. Hydrogen, therefore, as the scientists point out, becomes a key energy link in building up complex organic substances, not only in chemistry but in life itself.

Mindful of its portent, scientists are vigorously endeavoring to solve more of the mysteries of hydrogen. They are hopeful of success, for the sounds of space are continually whispering new scientific data and cosmological news from all over the heavens.

ELEMENTS OF THE ATMOSPHERE

Physicists explain that four elements: oxygen, nitrogen, helium, and hydrogen play important roles at different heights in the earth's atmosphere. They note that the density of each element falls off with the altitude because of gravity, but the lighter elements decrease less rapidly than do the heavier. Therefore, the relative composition of the atmosphere varies as a function of height. And the decline in density of each element depends on temperature; hence, knowledge of the temperatures at different altitudes is important.

Schoolboys are told that at the earth's surface the air consists of 78 per cent molecular nitrogen and 21 per cent oxygen; the remainder is carbon dioxide and other gases. Higher, somewhere between 1,500 and 3,000 kilometers, in the region known as the magnetosphere, hydrogen predominates. This is an expanse that fascinates scientists, challenged by its mysteries. They would like to know how the solar wind interacts with the magnetosphere. Do the lines of force of the earth's magnetic field at high altitudes return to the opposite pole of the earth or travel out into space to become associated with an interplanetary magnetic field?

Without magnetism, it is asked, could science account for the filamentary shapes of clouds, dust, and gas, or the majestic spiral structures of the galaxies? Little wonder that in the Space Age the new science of magneto-hydrodynamics, which relates to interactions between magnetic fields and electrified liquids and gases, has assumed a vital importance in theoretical astronomy, i.e., in considering the actual creation of planets, stars, and galaxies.

RADIO TELESCOPES

For the answers to such questions and many others, scientists and particularly astronomers have turned to new and more powerful radio-radar telescopes.

"Our 'window' for looking at the universe has opened wide," reports Harvard University, where for 200 years astronomers have studied the stars and planets through telescope, spectroscope, and photography. "Until the last decade, nearly everything we knew about the planets, stars, and galaxies came to us borne on light waves in the narrow, visible portion of the spectrum. Astronomers have likened the visible region of the electromagnetic spectrum—which stretches from short-wave gamma rays at one end and the long radio waves at the other—to a tiny window looking out on the universe. Most of the revolution in modern astronomy stems from the design of instruments that have opened new 'windows' in the electromagnetic spectrum.

"Many of these telescopes, cameras, radiometers, spectrometers, and spectrographs have been built, and others are being built, by astronomers at the Harvard College Observatory and the closely associated Smithsonian Astrophysical Observatory. The laboratories on Observatory Hill have become modern electronic workshops. . . . Although thermal photography is not new, the Harvard College Observatory has built one of the most sensitive instruments in existence for making thermal 'pictures' of the moon. It is about the size of a TV camera, and is called a radiation pyrometer.

"The science of radio astronomy, which has added much to our knowledge of the universe through radio telescopes, proves the galaxies in the universe to be a hissingly noisy chorus. . . . Astronomers suspect that the ionosphere itself may be a radio transmitter, and the Harvard astronomers hope to confirm this. They are also trying to find out how radio noise from the earth and its ionosphere fluctuates between day and night and from latitude to latitude, and how such things as sun-spot activity and solar flares influence its intensity. . . . The Harvard 60-foot radio telescope listens to 21-centimeter waves, the wavelength emissions from neutral hydrogen gas."

"Windows" in West Virginia Five radio telescopes dominate the landscape in Deer Park Valley, Green Bank, West Virginia, where the National Radio Astronomy Observatory (NRAO) is operated by Associated Universities, Inc., under contract with the National Science Foundation.

The oldest of these "windows on the universe" was first used in 1959. A 300-foot transit telescope recognized as the largest movable radio telescope in the world was completed in 1962 at a cost of about

$1 million. Its numerous assignments include probes of hydrogen's 21-centimeter "signals" and measurement of temperatures on the planets through a study of the spectrum of radio wavelengths.

A 140-foot telescope completed in 1965 cost $14 million and took seven years to design and construct. As the largest of equatorially mounted radio telescopes it is especially designed for use at short-wave lengths, and has been used successfully at a wavelength of 2 centimeters.

A portable 42-foot telescope is erected on a trailer so that it can be located at various sites near Green Bank to form a three-element interferometer. When two or more radio telescopes are connected they form what is known as one or more interferometers. Astronomers explain that such groups of telescopes make it possible to measure portions of the heavens to a very high accuracy. Also, if a part of the sky is observed for a long time, the results can be turned into a detailed map, very much as an optical telescope can photograph a patch of sky. The interferometer system at Green Bank comprises three steerable 85-foot telescopes, known as a synthetic antenna. With such a system of interconnected telescopes the astronomers find it possible to gain results as complete in detail as would be obtained by a single telescope 5,000 feet in diameter. To achieve such a "map" an area of the sky is scanned for many hours and subsequently the observations are interpreted by a computer.

Supplementing the Green Bank facilities, a 36-foot millimeter-wave telescope is located on Kitt Peak near Tucson, Arizona. The 6,000-foot mountain is the site of the Kitt Peak National Observatory, which is the optical-astronomy counterpart of the National Radio Observatory. This high, dry location was chosen since much of radio-wave absorption is caused by water vapor, the effects of which are reduced at the Arizona peak.

To help astronomers "take many more great steps ahead in their science," the National Radio Astronomy Observatory proposes a very-large-array (VLA) of radio telescopes. Plans specify a large number of antennas spread over a 20-mile area in the southwestern part of the United States. Regions in the sky can then be mapped in as much detail at radio wavelengths as can be obtained by optical telescopes. The radio astronomers using radio waves—which are akin to light waves differing only in their wavelengths—aim to match or surpass the measurements made by optical astronomers who use light waves.

Michigan Observatories Celestial investigations conducted by the University of Michigan Observatories, Ann Arbor, extend deep into solar space. Mercury, Mars, Venus, and Saturn have been continuously under surveillance by radio telescopes in attempts to improve measurements of the surface temperatures. The astronomers say that perhaps the most outstanding result has been determination of a Venus temperature of 675 degrees K (Kelvin absolute scale of temperature) equal to 756 degrees Fahrenheit. Surface temperature of Mars was indicated to be 182 degrees K (−132 degrees F).

In 1965, the Michigan scientists made a noteworthy discovery through measurements regarded as the first to establish variations in radio emissions from quasi stellar sources. And in 1966 it was found that those radio emissions showed variable polarization.

Illinois' Unique Telescope In November 1962, the University of Illinois dedicated a unique wire-mesh trough—a curved reflector— the size of five football fields. It is 600 feet long, 400 feet wide, equivalent in size to a 400-foot-diameter parabolic reflector. The mesh structure is laid on the ground in a steep-sided gully near the Vermilion River. Four 153-foot wooden towers along the center of the trough hold a catwalk on which a helical antenna array is mounted above the reflector. Embarking on decade-long investigations to map radio emissions from space, the Illinois astronomers plan to compile a catalogue of extragalactic radio sources by observations on 611 megacycles (49 centimeter wavelengths) at which the installation operates.

A new telescope went into operation at the Vermilion River Observatory in 1969. The astronomers describe it as being in the form of a paraboloid of revolution 120 feet in diameter, steerable in two coordinates so as to be able to track radio sources across the sky, in contrast to the 400-foot parabolic cylinder which is a meridian-transit instrument. It is used for radio spectroscopy of the galaxy and for other specialized radio astronomical observations, in some cases in conjunction with the 400-foot telescope.

Arecibo's Big "Ear" The world's largest radar-radio telescope, at Arecibo, Puerto Rico, was dedicated on November 1, 1963, to help make clear "the grand architecture of the universe." Taking advantage of a deep natural depression in the mountainous "sinkhole" terrain, a huge bowl was gouged as the seat of a 1,000-foot reflector—a dish antenna of galvanized steel mesh that provides a reflecting area of about 18.5 acres. Cables strung from three lofty concrete towers atop mountain peaks support the reflector, from which high-powered pulses beam into space and echo back.

Designed as a flexible tool for observation of the terrestrial ionosphere and for exploring the moon and planets, the telescope also can make passive observations of objects close to the edge of the observable universe. The $9.5-million instrument, conceived and constructed by Professor William E. Gordon while at Cornell University, is sponsored and financed by the U.S. Advanced Research Projects Agency and the National Science Foundation through the U.S. Air Force Office of Scientific Research.

Much insight has been gained about the ionosphere, including the discovery of vibrations at the plasma frequency and magnetic-line guidance of photoelectrons from the opposite hemisphere. Surfaces of the moon and Venus have been explored using radio wavelengths which partially penetrate the terrain. Measurements of the rotation of Venus and discovery of an unexpected 3/2-synchronous rotation of Mercury have been made. Some topics of research in radio astronomy have been the twinkling caused by solar corona, the inexplicably energetic emissions of quasars, and the highly periodic emissions of pulsars.

The Eyes of Texas Millimeter wavelengths represent the transition region between conventional radio astronomy and optical astronomy. Research in this transition frequency region is conducted with radio telescopes such as that installed at the Millimeter Wave Observatory of the University of Texas located on Mount Locke in West Texas. The 16-foot parabolic reflector made of invar—a steel alloy—is housed in an astrodome similar to the domes used for weather protection of optical telescopes. The dome rotates on rubber-tired wheels to align the shutter opening with the direction the telescope points so that it can scan any part of the sky.

Operation in the millimeter radio spectrum permits investigations previously considered inaccessible and impossible by telescopes attuned to "see" in the centimeter or meter wavelength regions. Since the Texas antenna is designed to make the telescope usable to wavelengths as short as 2 millimeters, it is expected to add a new dimension of knowledge of the emission characteristics of the solar system and galaxies. Scientists at the site report that valuable data on the moon, planets, and "radio" stars including pulsars are being taken at millimeter wavelengths. Solar observations are said to be particularly fruitful since greater depth into the sun can be seen with millimeter waves than with any other wavelengths.

Looking Out from California   The University of California operates a Radio Astronomy Laboratory in Hat Creek Valley in northern California. This $1.2-million installation, supported in part by the Office of Naval Research, features two radio telescopes, one with a dish antenna 33 feet in diameter, the other 85 feet. Both are designed to detect radio waves from the cosmos far beyond the range of the 200-inch optical telescope on Mount Palomar, California, which peers 2,000 million light-years, or 12,000 billion billion miles, into space.

The facility is expected to shed new knowledge of the Milky Way From a study of the distribution and motions of its gaseous material, the radio astronomers hope to learn more about the structure and dynamics of the galaxy. A multiplicity of mysteries exists in the Milky Way; and it will be centuries, if ever, before man even with the most powerful telescopes can hope to solve centuries-old conundrums about the stars and universe.

Another large radio telescope, built by the Stanford Research Institute with U.S. Air Force sponsorship, scans the skies from Stanford University, Palo Alto, California. It has a parabolic steel and aluminum 150-foot dish antenna and a 400-kilowatt radar installation for exploration of the sun, moon, and planets as well as interplanetary gases.

Signals from Jupiter   Twin 90-foot radio telescopes at California Institute of Technology's Owens Valley Observatory when pointed toward Jupiter are reported to have intercepted clues that the big planet is encircled by a powerful magnetic field. It is described as doughnut-shaped like the earth's Van Allen belt of charged particles trapped by the earth's magnetic field. The Jovian magnetic halo may be 300,000 miles in diameter and 80,000 miles thick, according to the radio observations. Does Jupiter have magnetic poles like the earth? The radio astronomers are inclined to believe that it does, and, further, they see evidence that the radio outbursts may be caused by auroras, thus casting some doubt on various other theories.

Radar astronomers in the Soviet Union announced that during the autumn of 1963 radar echoes were received from Jupiter when 370 million miles from the earth; it took one hour and six minutes for the signals to make the round trip. Echoes were also reported from Mars over a distance of 62 million miles.

Discovery by astronomers at Carnegie Institution of powerful long-wave radio "storm" signals from Jupiter dumbfounded the scientific world. Are such signals created by a non-thermal process near the surface of Jupiter, or do they originate somewhere in space around it?

Astronomers point out that the intensity of the Jovian signals indicates that *if* they were produced by thermal processes, Jupiter would be nine times hotter and therefore much brighter than the sun. But since that is obviously not the case, they wonder if non-thermal processes such as radiation of particles accelerated magnetically might be responsible. It is a mystery still to be solved.

Britain's pre-eminence in radio astronomy and tracking was gained by a 250-foot dish antenna at Jodrell Bank, which for many years was the largest steerable dish in existence—and it was quick to spot missiles and satellites. Experiments at this site are reported to reveal that all the planets are one foot a mile farther from the earth than previously calculated. For example, the distance between Mars and the earth, about 35 million miles at their nearest point and 63 million miles at the farthest, is now known within approximately 5,000 miles instead of previous estimates of 50,000 miles. And 2,000 miles have been added to the mean distance between the earth and sun—about 93 million miles. As a result, space scientists, mapping the orbits of planet-bound rockets, have revised their distance charts.

The Leiden Observatory, at Dwingeloo in the Netherlands, put an 82-foot dish antenna into operation in 1956, and that size is rated to be just about right for study of 21-centimeter waves.

The Australian National Radio Astronomy Observatory, at Parkes, New South Wales, features a 210-foot steerable radio tele-scope. Its planned investigations in a range estimated at about one billion light years, or six billion trillion miles, are described as numerous and diverse. Observations are made at wavelengths of 11, 21, 30, 75, and 220 centimeters. Part of the mission is to map the spiral arms of the Milky Way and similar features in the Clouds of Magellan and the Andromeda nebula.

Sensitivity of this telescope is so sharp that it is expected to map with hitherto unequaled resolving power, or precision, the structure of the galaxy in which the sun is one star in billions. Soon after it went into operation three times as many new sources of radio "noise" were detected in a given part of the heavens as had previously been known. Particular locations in the sky are said to have been pinpointed with such remarkable accuracy that there seems no doubt that exploration to the edge of the observable universe will ensue.

Since the Australian "dish" went into full operation early in 1961, it was put to work to study magnetic fields in distant galaxies, exposing, as the scientists declare, "a new horizon of opportunity to test existing theories and create new ones." From the radio astronomers at Sydney have come reports of evidence that magnetic fields exist within the radio source Centaurus A, about 13 million light-years away.

The Molonglo Radio Observatory 30 miles from Canberra features a giant Mills Cross, named after the inventor Professor Bernard Y. Mills, of the University of Sydney's School of Physics. The cross, comprising an array of 600 tons of steel framework and 22 acres of fine wire mesh, has two mile-long arms to intercept galactic radio signals.

In the Soviet Union, near Serpukhov, 55 miles south of Moscow, the Lebedev Institute of Physics of the Soviet Academy of Sciences operates a radio telescope featuring a gargantuan antenna system spread across 20 acres. Modeled after the antenna arranged in the form of a giant cross, designed by Dr. Mills of Australia, the telescope is wired to "see" emissions from extremely small and concentrated areas in the heavens visible to both arms of the cross. With the aim controlled electronically, two large dish antennas gaze on outer space by intercepting emissions on wavelengths as short as 8 millimeters, as well as various wavelengths characteristic of temperature-generating radiations from the moon and planets.

These radio observatories, about 90 throughout the world, have tremendous range, which encourages scientists to believe that they will find the answers to what they term cosmological questions: How big is the universe? When did it start? What is its shape? How was it created?

SEEING UNSUSPECTED THINGS

The moon is relatively easy to "touch," since it is only about 240 thousand miles away. The sun is the nearest star to the earth, and Alpha Centauri is second; it is 4.3 light-years distant, or 26 million million miles. That means it is 8,000 years away for a 100-mile per second rocket. A spacecraft traveling 200 miles per second would be scheduled to reach Alpha Centauri in 4,000 years. Nevertheless, it is within range of radio observation, estimated to cover five billion light-years.

Astronomers explain that if any galaxies are farther away their light can never reach the earth, because the distance is increasing faster than light can bridge the gap. Therefore, according to astrophysicists, the horizon, or edge of the universe, as far as man's vision is concerned, is put at about five billion light-years. Within that range they report the firmament is ablaze with mysterious radio noises, neither gravitational nor nuclear, but primarily electrical and magnetic, causing them to exclaim: "We are seeing things no one ever suspected!"

Already it has been discovered that the various celestial transmissions detected by radio astronomy are from objects other than the stars. For example, two sorts of radio waves come from the moon—waves reflected by it from the sun and thermal radiations emitted by the moon itself.

The "brightest" radio source, other than the sun, is reported to be Cassiopeia, 10,000 light-years away. Cygnus A, pictured as two colliding galaxies, 270 million light-years away, also is rated as a strong radio source, based upon the intensity of the radio emission with which it showers the earth. The Milky Way, according to astronomers, has as its radio counterpart a diffuse ribbon from the northern to the southern horizon; but as "seen" by radio, it increases in brightness along its length, reaching a peak of brilliancy in the constellation Sagittarius, which marks the center of the earth's galaxy. It is pictured as a giant cosmic wheel 1,000 miles in diameter containing a swarm of stars, with the sun as a speck of dust on a spoke of the wheel. And beyond the earth's galaxy, astronomers say there are at least 100 million more such galactic systems.

GAMMA RAY TELESCOPE

Anxious to learn more about gamma rays[1] and to obtain new clues for solving the mystery of their origin and distribution throughout space, NASA on April 27, 1961, launched a 95-pound satellite equipped with a special telescope designed to measure the previously undetectable gamma rays that streak through the universe at the speed of light.

This astronomical observatory, the thirty-ninth satellite launched by the United States, was developed and built at Massachusetts Institute of Technology. Shaped like an old-time street lamp, it was designed to tumble end over end ten times a minute, while orbiting from 310 to 1,100 miles above the earth and scanning a portion of the heavens every six minutes. The heart of the gamma ray telescope is described as "a sandwich of crystal layers composed of sodium iodide and cesium iodide." As gamma rays strike these crystals, an electron and a positron (a positively charged electron) are emitted. These particles then react with the crystals to produce flashes of light that are monitored by a scintillation counter, the reaction of which is radioed to earth. Scientists point to this as a significant new chapter in astronomy, since it was the first attempt to detect and measure gamma rays before they become lost in the babel of radiation that impinges upon the earth's upper atmosphere.

Evidence sent back by this satellite, Explorer 11, from its gamma ray telescope was evaluated by scientists as challenging the theory that the universe is in a steady state. It seemed to support the big bang, or evolutionary, theory that the universe was born with a gigantic primeval explosion. Preliminary findings made by the tele-

[1] Gamma rays are described as the shortest and hardest-hitting waves that come out of the transformation of a mass of disintegrating atoms into energy. They are not electrically charged and therefore are not deflected by magnetic fields but travel in straight lines; hence the direction from which they come can easily be determined.

scope were reported to reveal a lack of powerful gamma rays as would be created when matter and antimatter collide and annihilate each other.

Dr. Robert Jastrow, as chief of the theoretical division of NASA, in an address before the National Rocket Club said that the level of gamma radiation is so low as to "rule out one version of the steady state of cosmology which holds that matter and antimatter are being created simultaneously." Evidence from the satellite was analyzed as indicating that antimatter is not being created in the far reaches of the universe. If not, then scientists say that probably matter is not being created, as postulated by the steady-state theory on the formation of the universe.

INTERPRETING SOUNDS

While the universe seethes with electromagnetic jargon that defies interpretation, the radio astronomers find themselves in much the same predicament as were physicians when an electronic stethoscope was developed. When they applied it to the heart and other human organs, so many new and strange sounds were heard that they gave it up in confusion and returned to the standard stethoscope which produced sounds they had learned to understand.

So too with the electron microscope. It reveals so many new things—never before seen—in the submicroscopic world that it will require years of study to interpret the full significance of the infinitesimal.

"As the universe is boundless one way towards the great, so it is equally boundless the other way towards the small," said Sir Oliver Heaviside,[2] "and important events may arise from what is going on inside atoms and again in the inside of the electrons. There is no energetic difficulty. Large amounts of energy may be condensed by reason of great forces at small distances. How electrons are made has not yet been discovered. From the atom to the electron is a great step but it is not finality."

Now, years later, electron microscopists and radio astronomers are

[2] In Volume III of his *Electromagnetic Theory*, published in 1912.

probing the boundless universe—one way toward the small, one way toward the great.

As the resolving power of the electron microscope delves deeper into the unseen domain of the infinitesimal, and the radio telescopes peer further into the reaches of space, the problems and challenges of interpretation will multiply. To classify and catalogue the tiny creatures brought into view and magnified by the electron microscope; to decode and decipher, translate and understand the lapping of the waves in the immense cosmic ocean, is a gigantic task for many a generation.

It was Marconi who said, "The more man bends the phenomena of the universe to his will and the more he discovers, the more he will find to discover. Because of this he will realize more and more the infinity of the Infinite."[3]

"There is something frightening about the universe," said Nikola Tesla,[4] "when we consider that only our senses of sound and sight make it beautiful. Just think, the universe is darker than the darkest ink; colder than the coldest ice, and more silent than a silent tomb with all the bodies rushing through it at terrific speeds. What an awe-inspiring picture, isn't it? Yet it is our brain that gives merely a physical impression. Sight and sound are the only avenues through which we can perceive it all. Often I have wondered if there is another sense which we have failed to discover. I'm afraid not," he added after some hesitation in thought. "If there were, we might learn more about the universe.

"I do not believe that matter and energy are interchangeable, any more than is the body and soul. There is just so much matter in the universe and it cannot be destroyed. As I see life on this planet, there is no individuality. It may sound ridiculous to say so, but I believe each person is but a wave passing through space, ever-changing from minute to minute as it travels along, finally, someday, just becoming dissolved."

Little wonder that those studying space travel humbly declare: Our first voyages into space are but an infinitesimal step into the 600 billion billion miles of the invisible eternity surrounding us.

[3] Interviewed by the author, New York, 1927.
[4] Interviewed by the author in April 1934.

XVII   Tests of Time

> The world into which we were born is
> gone; we have little or no idea of the
> world into which our children may grow
> to maturity. It is this rate of change, even
> more than the change itself, that I see as
> the dominant fact of our time.
> —*Julius A. Stratton*

IRONICALLY, EVERY OFFSPRING OF WIRELESS CREATED COMPETITION
for its parent and threatened the inventions and systems from which it
sprang. Everything in the science of communications is subjected to
the test of survival of the fittest.

In every instance, the acid test is: What can this invention or
system do to improve or surpass an existing service; what does it offer
that is new, more efficient, and more economical? To what further
advances is it a key?

To satisfy this rigid formula all inventions and services have had to
maintain pace with progress and quickly adapt themselves to new
developments and change—or else suffer the penalty of extinction.

Everything old in radio and electronics, no matter how miraculous
or dramatic in the past, is challenged to compete with the new to
survive, or else pass into the graveyard of obsolescence as did the
spark gaps, the arcs, the alternators, and so many other instruments
that served their day and vanished into antiquity.

Those who failed to realize the potentialities of wireless were blind
to the new and far-flung services it could perform—better and more
economically.

WIRELESS *vs.* CABLES

Sir John Wolfe-Barry, presiding at a meeting of stockholders of the Western Telegraph Company, Limited, in London on October 30, 1907, said,[1] "I do not look upon any system of wireless telegraphy as a serious competitor with our cables." He did not stop to evaluate what wireless could do that the cable could not. In fact, he brushed wireless aside because of what he called its "fundamental difficulties and imperfections," giving little thought to the fact that science and engineering might overcome them. In wireless, he saw too much "uncertainty in transmission and lack of secrecy" as well as "the confusion that must occur when numerous messages are not conveyed by a direct conductor," such as the cable or land line. He explained that wireless messages were "thrown with great violence into the ether into which they radiate in all directions, to be disentangled and interpreted on the other side of the globe."

In the cables he saw a certainty of transmission, while wireless had to encounter many atmospheric disturbances which made the task even more formidable. He perceived great opportunities for error and the constant demand for repetition of doubtful signals. And he concluded: "If the order of discovery had been reversed and the cables had come second in point of date to a wireless system, it would have been universally recognized that cables were a fundamental and enormous improvement, eliminating the difficulties under which a wireless system must labor." Such was one outlook for wireless in 1907. Nevertheless, as early as 1901, *The New York Times* pointed out:

. . . The thing he (Marconi) is attempting to do would be almost transforming in its effect upon the social life, the business and political relations of the peoples of the earth. . . . The electric telegraph, in the form of ocean cables, was a great step in advance. The sending of messages without wires through natural media of communication will be a still longer and more wonderful advance, if it shall prove that the art can be perfected and made practicable up to the measure of present confident predictions. Everything depends upon that. The cables are too slow and too costly in these modern times. . . .

[1] *The London Electrician,* November 1, 1907.

It is a well-known story that wireless measured up to the confident predictions made at the turn of the century. Seventy-odd years have put wireless to the test, and it has survived the vicissitudes of time because it kept pace with progress, ever-changing and always up to date on the thousands of channels that crisscross the hemispheres.

In the wireless *vs.* cable controversy, no thought seemed to exist that someday the telephone might need them both and put new demands upon their facilities. The telephone in those days was looked upon as a domestic affair. Now, however, telephone traffic to overseas points is a vital factor in the life and expansion of radio, cable, and commerce as a whole. Ever since 1950, when there were about one million overseas phone calls, growth has accelerated. Expansion continues at about 20 per cent every year, with increased demands for more facilities to handle high-speed data transmission and other special services such as overseas television.

Now, even with a satellite system to relieve the load by providing many hundreds of voice channels, communication experts declare that the undersea cable certainly will remain an integral part of overseas communications for decades to come. Additional cable systems are under construction, despite the fact that satellite systems hold promise of lower per circuit cost where large numbers of circuits are needed.

The Bell System's fifth transatlantic cable, a 3,500-mile link with a 720-channel capacity, was begun in 1969 between Green Hill, Rhode Island, and San Fernando, Spain. It includes a radio relay link to Portugal, also a radio relay and Mediterranean cable link to Italy, all of which were planned for operation in 1970.

COMPUTERIZED TELEGRAPHY

Estimating that its world-wide commercial traffic will exceed 300 million words by 1970, RCA in 1964 demonstrated a computerized system described as "the fastest, most accurate and versatile international public telegraph system ever developed." it replaces perforated tape for automatically transmitting and receiving messages, and is capable of handling any mixture of high-frequency radio channels, submarine cables, radio satellite, or wireline channels.

A single high-speed computer is capable of processing 2.5 million characters, or 400,000 words per second. In transmitting and receiving messages to and from approximately 70 countries, the computer can identify in a fraction of a second any one of 3,000 cities, states, and countries in three languages, and any one of 9,000 internationally registered coded addresses. After the computer decides where, when, and how the message is to go, and in what order, the message is directed to output buffers which convert the high internal speed of the computer to the relatively low speed (60 to 100 words a minute) of the local teleprinter equipment and the overseas transmitting channels. The computer also examines automatically all messages for the required accounting and customers' billing information.

NEW INSTRUMENTS FOR OLD

Man-made instruments may be relatively short-lived, and at times it may appear that only the basic principles remain in force. A typical case in point is the massive high-frequency alternator, which during World War I proved its worth in international communications as well as the strategic value of wireless. It made radio history in both an engineering and a political sense.

At the close of the war, the Marconi Wireless Telegraph Company of America,[2] an offshoot of the British Marconi Company, was the only company in a position to handle commercial transatlantic radio communication. Negotiations for the transfer of patent rights as well as alternators from the General Electric Company to the Marconi Company were resumed in 1919, having been interrupted by the war. Fears were expressed in Government circles that American overseas communication service would be under foreign control. It was suggested that negotiations be suspended until after discussions with the U.S. Navy Department. A plan was developed to form a new American company to acquire the assets of the American Marconi Company. As a result, the Radio Corporation of America was formed in October 1919, with the objective of giving the United States preeminence in international radio communications, independent of

[2] Organized, November 22, 1899.

foreign control. Radio Central was established in 1921 as a world center of communications at Rocky Point and Riverhead, Long Island, with high-frequency 200-kilowatt alternators as the powerful driving force. Stations taken over from the Marconi Company by the Government during the war were turned over to RCA for foreign communication service.

The dynamo-like alternators doomed the high-power spark and arc transmitters. But the alternator, like everything else in radio, ran into competition in 1922 when development of powerful electron tubes and shortwaves threatened to send it into oblivion along with lofty antenna towers. By World War II the alternator was an antique. A veteran wireless man nostalgically suggested that rather than scrap the alternators as junk, they be set up to mark historic radio sites in much the same way obsolete cannons and relics of war are used in parks and public squares. No one among modern communication men was interested.

The interests of today are not the interests of yesterday; neither is the radio of today the wireless of yesteryear. Its services are new, and much of its operation is swift and automatic in handling millions of words day and night, year after year.

RADIO vs. PHONOGRAPH

As the cables frowned upon wireless, so the phonograph laughed off radio as a joke, a passing fad. The phonograph had the great artists and their music on disks uncontaminated by atmospheric noises such as static. The phonograph offered "music you want when you want it." Radio could not boast that. But the phonograph people overlooked the fact that radio could do something the phonograph could not. While Caruso, Galli-Curci, and others went round and round on disks, radio went around the country and around the world into millions of homes. Here was a new opportunity for the phonograph to reach into countless homes to publicize its music and popularize its artists. How could the hand-wound phonograph of those days compete with radio's fascination and instantaneous mass appeal? But the phonograph wanted no part of it; its popularity declined. Finally, in 1929, the Victor Talking Machine Company was acquired by RCA, which electronized the Victrola, popularized it, and spread its entertainment value on an unprecedented scale. The record business flourished and prospered.

TV VS. RADIO

After 25 years of growth and prosperity, radio broadcasting ran into television competition—a new service operating in the same medium and offering the extra feature of pictures in motion.

Since the early twenties, radio met challenge after challenge—always changing, always confronting something new and improved. But television is something bigger. With its dual appeal to eye and ear, it does not depend upon the magic of illusion that accounted for the long run of "soap operas" and many other radio programs. Television gives the imagination a rest and makes the mind lazy when compared with the way that radio toyed with it by playing upon the imagination. Television is more realistic; the eye is not fooled as easily as the ear. Seeing is believing, not make-believe.

Television, an offspring of radio, offered a new service—sound and sight combined. It offered much that radio did, but with intensified mass appeal. Then why, some asked, will radio broadcasting survive? Would not television capture the audience not only from radio but from sports events upon which the cameras focus?

"No," said one fellow, "to stay home and watch a prize fight or baseball game on TV is like staying home looking at a picture of your best girl instead of being with her!"

Nevertheless, television's effect on radio was revolutionary. Radio had to change its ways and concentrate on what service it could perform that television could not—for example, in automobiles and in small portable form. News, spot announcements, and local advertising were its strong forte for survival. And the transistor that made radios pocket-sized enhanced its usefulness. Be that as it may, radio broadcasting slipped in the fifties as television gained at radio's expense. The disembodied voice alone found it increasingly difficult to compete with a voice and a smile.

The younger generation wondered how its parents could ever have stayed home night after night following Amos 'n' Andy and the invisible cavalcade of radio; it was equally difficult to understand how they tolerated silent movies. Television was different.

Dire predictions were heard that the impact of television on radio would be like the automobile's effect on the horse-and-buggy. The tests of time and competition had caught up with radio.

And while battling to survive against television, AM radio faced another competitor in FM. In the thirties, it was called an inventor's dream, and FM was referred to as the symbol of a frustrated medium. Nevertheless, FM continued to grow, despite the fact that it ran head-on into television in the post-war period. Sales of FM home receivers steadily climbed several million a year.

Ask the FM advocate why, and he replies that FM is the story of the better mouse trap. Tonal excellence and freedom from static as well as other extraneous noises qualify it to survive and expand in competition with the standard broadcasts of AM. Indeed, FM supplements and strengthens radio's appeal in competition with television.

TV *vs.* HOLLYWOOD

Television, endowed with immediacy and with marked ability to handle films, worried filmland too. Was it a competitor or a customer? Some caught the vision of a new specter of obsolescence hovering over the screen. Here was a new medium opening a vast new theatre for motion pictures by the fireside. Again the question: What can TV do that Hollywood cannot? It could reach many millions of people simultaneously with entertainment in their homes.

"Motion pictures, in my judgment, will be the sturdy backbone of television," said Eric Johnston, president of the Motion Picture Association in 1948.[3] "I believe that a great spurt in film production and forward strides in picture-making techniques are inevitable."

This, of course, was not the first time that movies had been called upon to meet tests of survival; but television seemed to be somewhat out of their control, causing most of the moguls of screenland to shudder at the prospects. They remembered when predictions were heard in the twenties that sound on film was coming; that experimenters were intent upon turning the silent pictures into talkies. But that was within the theatre and under their direct jurisdiction.

They remembered, too, how in early demonstrations amplified

[3] Hollywood Reporter booklet on *Television and What the Motion Picture Industry Is Thinking and Doing About It*, November 1948.

sound was unnatural and threatened to turn the peace and quite of the silent theatre into a noisy place. Who would tolerate blasts of sound detracting from enjoyment of the picture? It was preposterous! All such thoughts began to change, however, when, in 1927, Warner Brothers presented *The Jazz Singer,* featuring Al Jolson. Almost overnight the silent actor became a thing of the past, and the film industry, aided by science, entered a new era of popularity and expansion.

Now, television, hungry for films, has found a direct relationship with filmland which can be cultivated for a new era of prosperity. As television becomes more and more film- and tape-minded, Hollywood finds that it has the stars, techniques, and facilities to make productions for TV; also it has a new market for films long ago relegated to the files.

Moreover, Hollywood can concentrate on feature pictures and spectacle-type films such as *Ben Hur,* limited in exhibition to the theatre screens. Hope lurks also in the old assertion that man is a gregarious creature and does not like to sit home all the time watching a 21-inch fluorescent screen. Surely, it is reasoned, feature films such as *Cimarron* are more interesting on giant theatre screens, so why worry about the relatively small video screens? Wall-sized TV for the home, predicted for the future, however, gives filmland something else to think about.

PAY TV

Both television and the films, however, have mutual interest in watching the results of closed-circuit pay-TV tests in which programs are "piped" into homes over wires to coin-box TV sets, or radioed into the home as a coded signal. The pictures and sound are scrambled at the TV transmitter and the coin box in the home unscrambles.

If pay TV becomes successful on a big scale, free television through the air would no doubt lose some of its audience. At the same time, pay TV would compete with theatres for patronage and might even bid for first-run films; it could also carry advertising. Hollywood, on the other hand, might gain in that it would have a new outlet for films. Talent performing on television might move over to tollcasting.

Then major sports events such as the world series and championship fights might find it more lucrative to go on pay TV, not only to be "piped" into homes but into theatres and auditoriums throughout the country.

Practical tests will demonstrate, and time will reveal whether or not pay TV has widespread popular appeal and will determine how the subscription idea fits into the entertainment picture in competition with the immediacy and universality of free broadcasting. In any event, the first 25 years of television disclosed no signs that the movie theatres were to be emptied, that Broadway would go dark, and that radio broadcasting would be silenced as a thing of the past.

CABLE TV

Cablecasting (CATV) when introduced between 1947 and 1950 was called "Community antenna television." The idea was to erect a master antenna atop a hill, mountain, or skyscraper to provide clear reception in areas shielded by mountains and buildings, or too far distant from stations to pick up proper signal strength, if received at all. Therefore, why not install a high antenna and rebroadcast the signals with maximum strength into the "dead spots" or valleys of the community? As an added advantage the central antenna could be served by microwave relay stations lined up across the country.

From that radio-relay concept emerged the plan to link homes to a control station that amplifies and passes the signals to a main cable extending through the community on poles or in conduits. From the main cable a coaxial cable link leads into the homes of subscribers for which they pay an installation fee and monthly service charge. In return they receive over-the-air broadcasts that are out of range of their rooftop or indoor antennas in addition to programs produced for cable transmission, not put on the air. And "ghosts" caused by reflection, or "echoes" of the signals from mountains or steel structures, are eliminated. The cable plugged into home TV sets gives the family more channels from which to select. While an over-the-air station on its single channel presents one program at a time, the cable can offer a variety of programs or services, 50 or more.

Cablecasting, estimated to be serving several million receivers in 1969, loomed as a form of pay TV. It opened new channels of entertainment and information that might range from first-run movies to sports and news as well as other classifications of programs from drama to education and politics. The prospects lead some to wonder if the day might come when pay TV and CATV would be wedded into a cable system.

In any event the future of CATV seemed destined to evolve through tests similar in many respects to those that radio broadcasting and television went through in Congress, the FCC, the Department of Justice, and the courts. Numerous technological, social, economic, and legal problems were to be solved.

Questions seeking answers concern: program content and sources, franchise ownership, interconnection of cable systems on a regional, national, and international scale, copyrights, commercial sponsorship, competition between cablecasters and over-the-air telecasters; also, should cable TV be declared a public utility; to what extent should restraints be imposed, if any; what regulatory concepts of over-the-air television are applicable to TV distributed over wires to subscribers?

Rules on CATV announced by the FCC in October 1969 freed cable TV from local restrictions on programming. Any type of programs including sports and other events as well as movies could be presented in competition with the regular commercial networks and stations. Commercials could be inserted during "natural breaks" in the programs. The regulations specified that cable systems with more than 3,500 subscribers would be required to originate some of their own programs as of January 1, 1971; and that equal time be afforded political candidates. Programs would be subject to sponsorship identification as on commercial stations.

COLOR *vs.* MONOCHROME

Color television became another competitive factor not only for radio and the movies but more so for black-and-white TV. In 1968, for the first time, color TV set sales exceeded black-and-white; 5.8 million color units, 5.5 million monochrome. There were 64.5 million monochrome sets in use and 20.1 million color receivers in the United States.

Those who once questioned whether color TV had a future overlooked all color offers that black and white lacks. As color glorifies the vision, so it glorifies television. It brings new life and realism—a new dimension—to the entertainment screen. It is a boon to advertisers and merchandisers. Color helps to identify products and vividly impresses them on the mind. Seen in color on the TV screen, package goods, whether a box of cereal or a can of soup, are more quickly spotted on the food store shelf. Recognition motivates sales.

Fortunately for television, Americans are color-minded, as evidenced by the color appeal of automobiles, refrigerators, kitchen appliances, telephones, and all sorts of household products. Color on TV not only enhances its entertainment value but extends its power as an advertising medium. Visual demonstrations in color are natural; they sell through the eye. How better to present and popularize a new red rose or make a golden cheese soufflé look delicious? Whoever saw a black and white strawberry, apple, sunflower, or dahlia?

The popularity of color photography and expansion of color advertising supported the faith and encouraged the hopes of the champions of color TV. It was declared that nothing scores like color; implementation of color in newspaper advertising was heralded as a bonanza.

Surveys reveal that the attitudes of readers toward products advertised in color are generally more favorable than those advertised in black and white. Evaluation by readers shows that color affects the senses and because of it, the product advertised is more readily visualized and the moods produced are cheerful and lively. Color, it is agreed, provides a powerful thrust for the creative talents of the advertiser.

Whatever print can claim for color so can television. It has appetizing aspects for all sorts of food products from baked goods to meats. It has "deep sell" that stimulates sales of everything from rugs to automobiles. Institutionally it paints a sharper image.

Just as the newspapers discovered the resounding, product-moving impact of color's emotional stimulus to the public so did television on an ever-increasing scale, as all-out industry interest and promotion got behind it. Upward went the sales curve. More and more programs were colorcast. Color sets were available in a wide variety of cabinet designs and as portables. The time came when a station not broadcasting color was out of step with progress and popular appeal. By 1969 the majority of commercial stations had color network capability; a high proportion could originate color live, on film or tape.

Until the home has color TV, the family will miss television at its best. The industry will continue to expand because color is an attraction that TV monochrome does not offer. And three-dimensional color pictures are in the offing.

AT THE WORLD'S FAIR

Television came to the 1964 New York World's Fair in color. Three hundred color receivers, an all-color mobile unit, and see-yourself cameras exhibited a quarter-century of progress since TV was launched as an industry on the same site in 1939.

When Franklin D. Roosevelt, as the first President to be televised, opened the 1939 fair about 200 small-screen TV sets were within range of the telecast. The pictures inspired an observation that "possibly within 20 years gazing by radio at a president, queen or king, even across the sea, will be no extraordinary scientific achievement." True, indeed, for when President Johnson opened the 1964 fair he was seen by many millions, and satellites were in the sky to relay scenes across the seas.

Things unknown and undreamed of in 1939 have gone into service. Flushing Meadows in 1939 saw no exhibits of communication satellites, radar, jet planes, rockets, atomic power, or astronauts in outer space. As President Johnson observed, "The reality has far outstripped the vision." Since then science has created the maser and laser, transistors, and semiconductors, solar batteries, microscopic circuits for miniature instruments, magnetic tapes, electronic memory units, and an array of computers. Picturephones enable users to see each other while they talk. Payphones—1,400 of them at the fair—featured the new touch-tone type in which buttons are pressed instead of dialing. Machines "talk" to machines. These and many other innovations such as coherent light promise to revolutionize communications in the next 20 years so that by 1990 the marvels of 1970—tested by time—will be vastly changed as are the wonders of 1939 in the seventies.

RADAR ADVANCES

Radar, too, will be under constant improvement. Operating on both radio and light beams, and abetted by the laser, it will make remarkable advances within its own scientific framework, rather than in competition with any foreseeable new service or system that will outperform it. As a tracking system at airports it determines a plane's direction, location, and altitude.

Radar is a service born of radio-television; it was fashioned to a great extent from their techniques. New devices, such as the magnetron and radarscope, were developed to perform a radically new service vital to the military, to shipping, and to aviation. And since there is no other system that accomplishes what radar does, it has the earmarks for survival. Indeed, a tremendous future is foreseen for it because of its link with the Space Age.

Amidst all of this activity in electronic communications, radiophoto also has met the test of time and continues to expand by virtue of the fact that as a service it performs parallel to its cousins, worldwide radiotelegraphy, radio broadcasting, and television. Radiophoto supplements the high-speed dots and dashes that spell out the radiograms; it supplements the teleprinters and other services that radio offers and does the job with dispatch and efficiency to merit continued existence along with phototelex.

As the science of wireless sired radiophotos, so it is the patriarch of all other electromagnetic communications in space. Radio broadcasting, television, radar, and interspatial signaling are all satellites from the launching pad of electronics for flights of progress that lead into the future.

How far each service or system goes depends upon the tests of time which determine the span of usefulness for all inventions and their applications. Basically these tests are the same in the 1970s as in the days of Marconi, Fessenden, and deForest. Sooner or later everything new has to measure up and prove its worth.

TESTS TO COME IN SPACE

What of the future; what next now that men have walked on the moon?

Test after test is certain to come as man seeks to develop and demonstrate his capability for interplanetary travel and to unlock ageold secrets of the universe. Lunar landings are not an end but the beginning of endless challenges to explore and to voyage in the heavens. Scientists are hopeful that some of the mystery of the 4.5-billion years of evolution of the moon and earth will be unraveled as astronauts continue to probe for information. No doubt many surprises and unpredictable circumstances lurk beyond the horizon but foreseeable steps into the future include:

1. A fleet of Apollos: lunar expeditions to touch down in different areas of the moon between 1970 and 1975.

2. Lunar exploration and construction of scientific bases on the moon; establishment of an earth-to-moon shuttle.

3. Manned space stations—orbiting laboratories—in earth-orbit to scrutinize weather, assay mineral resources, agriculture, forests, seas, and waterways; study geography and geology among other earthly activities on a global scale; and make interstellar observations uninhibited by the earth's atmosphere.

4. Exploration of Mars in 1971 by unmanned Mariner spacecrafts to orbit, not fly by. And in 1973 by Vikings—a Surveyor-type and a Mariner Orbiter to soft-land seeking to answer, "Is there life on Mars?"

5. Two Pioneer spacecraft to fly by Jupiter in the 1970s.

6. Man landing on Mars possibly by the end of the century.

NEW WORLDS TO CONQUER

Modern technology moves with swiftness. The time lapse between research and discovery to development for practical use is much shorter—much less than from wireless to television. Never before have so many scientists and engineers been at work in such superbly equipped laboratories. So rapid is the stream of new knowledge that techniques and instruments heralded as marvels only a few years ago already have served their time and are being replaced by a plethora of innovations demanded by the Space Age.

Any boy inclined to think there are no new worlds to conquer has but to look into space. Opportunity shines like the brightest star. Plasma physics and ion engines, solid-state physics, quantum electronics, and integrated electronics, computers, and data processing are intriguing fields that beckon talent on broad fronts of science, engineering, and industry. And there are all phases of radio, television, radar, telemetry, and radio astronomy as well as light-beam communications, the laser and optics. Bionics, medical electronics, nuclear reactors, thermoelectric power, and production of electricity by direct conversion from light, heat, and chemical energy are among the constellations of opportunity in the firmament of technology, invention, communications, and aerospace.

Science referred to in the past as "on the march" is now in swift flight. Man in orbit is on a new rendezvous with Nature, which, when probed, reveals secrets that lead to progress on earth. One has but to scan the heavens to behold the infinite mystery of the universe, and to realize the boundless opportunities that exist for all men and nations in the inexhaustible laboratory of space.

Index

Aerospace, 171–172
Air space, 171
Alaska Communications System, 290–291
Alcock, John, 23
Aldrin, Edwin E., Jr., 110, 136–147
Alexanderson, E. F. W., 30, 64, 66
Alouette satellite, 213
Alpha Centauri, 312
Alternator, high-frequency, 29–30, 320
AM (Amplitude modulation), 49–53, 147, 322
Amateur radio, 21–23, 31, 32, 200
American Optical Co., 252, 261
A.T.&T., 37–41, 47, 160, 174, 183, 288, 318
Anders, William A., 122–127
Andover, Maine, 175, 179
Angstrom unit, 249
Apollo missions: radio, TV, radar, telemetry, 117–149
    Apollo 6, 118
    7, test of redesigned capsule, 118–120
    8, circumlunar flight, 121–127
    9, lunar test flight, 127–130
    10, lunar rehearsal, 130–135
    11, moonlanding, 136–147
    12, moonlanding, 147–149
    Apollo Experiments Packages, EASEP, 145–146; ALSEP, 148
Appleton, E. V., 76
Arecibo Ionospheric Observatory, 307–308
Ariel satellite, 212, 213
Armstrong, Edwin H., 11–12, 31, 41, 49–53
Armstrong, Neil, 105–106, 136–147
Astro-electronics, 144–147, 152–153, 168–170

Astrogation, 87
Astro-monkey, (chimpanzee), 91, 219
Astronauts, 86–111, 112–135, 136–149; see also Cosmonauts
Astronomy, radio, 200–204, 209–212, 272–274, 295–297
Atmosphere, 94, 97–98, 140, 171, 211, 227, 229–230, 245, 271, 287–288, 290, 304
ATS-1 satellites, 130, 198–199
Audion, 30–32, 268
Australian National Radio Astronomy Observatory, 310–311
Autodin, 292–294
Automation, 58–59, 281–282, 292–293, 318–319
Aviation radar, 132–133, 327–328
Aviation radio, 23–27, 133
AWACS, airborne warning, 78

Baade, Walter, 296
Bain, Alexander, 55
Baird, John Logie, 64, 66
Bardeen, John, 53
Bar Harbor, Maine, see Otter Cliffs
Bean, Alan L., 147–149
Beaverbrook, Lord, 75
Bell Telephone Laboratories, 53, 66, 75, 156, 157, 176, 246–249, 300
Belyayev, Pavel I., 98–99
Bendix Corp., 173
Beregovi, Georgi T., 120
Big Bang theory, 300–301, 313–314
Binns, Jack, 13–15
Bionics, 274–275
Biosatellites, 218
Bloembergen, Nicolass, 247
BMEWS, 78, 291
Boeing Company, 169, 239
Borman, Frank, 101, 122–127

Boyden Observatory, 261
Branly, Edouard, 6, 8, 16
Brattin, Walter H., 53
Braun, Karl Ferdinand, 65, 67
Broadcasting, history and development, 28–54, 321–322
Brown, Sir Arthur Whitten, 23
Bykovsky, Valery F., 96
Byrd, Admiral Richard E., 26

Cable TV (CATV), 324–325
Cables, 183, 317–318
California Institute of Technology, 303, 309–310
Campbell-Swinton, A. A., 68
Canadian satellites, 213
Cannon, George C., 22
Cape Canaveral, 86; see also Cape Kennedy
Cape Kennedy, Space Center, 86–87
Cape Race, Newfoundland, 13–15
Carpathia, S.S., 13–15
Carpenter, M. Scott, 92
Carson, Arthur N., 262–263
Cassiopeia, 312
Cathode-ray tube, 74–75, 82
Cavendish, Henry, 302
Cernan, Eugene A., 106–107, 130–133
Chaffee, Joseph G., 157
Chaffee, Roger B., 111
Charyk, Joseph V., 161
Chelates, 264–265
Citizens radio, 290
Coherent light, 101, 249–250, 260
Coherer, 8–10, 11
Collins, Michael, 107–108, 136–147
Collins Radio Co., 157
Color TV, 70–72, 81, 127–135, 144, 148, 149, 177, 184, 198, 325–327
Columbia Broadcasting System, 70–72, 261–262
Communications Satellite Corp. (COMSAT), 160–161, 183, 188–189
Communications satellites, 174–190
Computerized telegraphy, 58–59
Computers, 58–59, 88, 92, 102, 109, 135, 138, 146, 194, 277–278, 281–282, 318–319
Computerized typesetting, 281–282
COMSAT, see Communications Satellite Corp.
Congress, 89, 91, 96, 115–116, 127, 143, 160–161

Conrad, Charles, Jr., 100, 109, 147–149
Conrad, Frank, 33–35
Coolidge, Calvin, 28, 41
Cooper, L. Gordon, Jr., 95–96, 100
Cornell University, 308
Cosmic geography, 295–297
Cosmic rays, 202, 286, 296
Cosmodogs, 150, 215
Cosmonauts, 86, 90, 93, 96–97, 111, 120–121
Cosmos, 95, 134, 154, 295–297
Cosmos satellites, 214–216
Cottam, Harold, 15
Courier 1B satellite, 151
CQD, 13, 14
Crimean Astrophysical Observatory, 243
Cripps, Sir Stafford, 76
Crookes, Sir William, 16, 67, 272
Crystal detector, 36, 41, 54
Cunningham, R. Walter, 118
Cygnus, A., 296, 312

Da Vinci, Leonardo, 167
Davis, H. P., 34–35
Dawson, Geoffrey, 48
DeForest, Lee, 21, 30–32, 35, 54, 64, 268
DEW Line, 78, 291
Digital (analogue) picture transmission, 227, 239, 278
Dolbear, Amos E., 16
DuBridge, Lee A., 97–98
DuMont, Allen B., 64
Dyson, Freeman J., 269

Early Bird satellite, 187–188
use during Gemini flights, 100–110
Eastman Kodak, 239
Echo balloons, 151, 156–157
Edison, Thomas A., 11, 16, 30
Edison effect, 30
Editor & Publisher, 44
Edward VIII, King, 46
Eglin Air Force Base, (radar), 172
Ehret, Cornelius D., 51
Einstein, Albert, 3, 17, 83, 246
Eisele, Donn F., 118
Eisenhower, Dwight D., 151, 156, 178
Electromagnetic waves, discovery of, 1–6
Electron, 9, 30, 314; see also Electronics

Electron microscope, 314
Electron tubes, 9, 30–31, 36, 54, 278, 320
Electronics, 9, 30, 53–54, 164, 170–172, 277–279, 280–285
Electronic Industries Assoc., 171–172
Elettra, yacht, 18
Elint, 170
Ellsworth, Charles B., 13–14
Eros satellite, 208, 259
Essa satellites, 196–197
Ether, 1–3, 256–257
European Space Research Organization, 214
Eurovision, 93
Evans Signal Laboratory, 84
Explorer satellites, 150, 209–211, 271, 313
Extra Vehicular Communications System (EVCS), 147

Facsimile, *see* Radiophoto
Fahy, Thomas P., 260
Farnsworth, Philo T., 64
Federal Communications Commission (FCC), 69–70, 71, 155–156, 160, 288, 290, 325
Feoktisov, Konstantin P., 98
Fessenden, Reginald A., 16, 28–30
Fleming, Sir Ambrose, 12, 16, 30
Florida, S.S., 13
FM (Frequency modulation), 49–53, 61, 147, 255, 322
Ford, laser experiments, 265
Foucault, Leon, 281
Fournier, M., 65
Frankfurter, Felix, 16–17, 72
French satellites, 213, 214
Frog's vision, 274–275

Gabor, Dennis, 261–262
Gagarin, Yuri A., 86, 111
Galvani, Luigi, 275
Gamma rays, 85, 201–202, 313–314
Gamma ray telescope, 201–202, 313–314
Garrett, C. G. B., 250
Gemini Project, 99–111
General Electric Co., 30, 36, 130, 148, 160, 255, 272, 276–277, 319
General Telephone & Electronics, 160, 266
Geoengineering, 164
Geophysics, 145–146, 164, 200–204, 210–211, 254–255

Geoscience electronics, 164
Glenn, John H., Jr., 91–92, 96, 191
Goddard Space Flight Center, 211
Godley, Paul, 21
Goldsmith, Alfred N., 45, 64, 71
Goldstone Tracking Station, 123, 131, 156, 222, 223, 230, 231, 297
Gordon, Richard F., Jr., 109–110, 147–149
Gordon, William E., 308
Graphology, space, 279
Gravity, 124, 132, 141, 152–153, 158, 211–212, 217, 240, 269
Green Bank, W.Va., *see* National Radio Astronomy Observatory
Grissom, Virgil I., 89–90, 99, 111
Ground Control Approach (GCA), 79–80
Grumman Aircraft, 129
Guided missiles, 18–20, 170
Gunn, J. B., 289
Gyroscopes, 280–281

Hals, Jorgen, 82–83
Hammond, John Hays, Jr., 18–20
Harvard College Observatory, 261, 298, 304–305
Harvard University, 247, 250, 304
Hawker, Harry G., 25
"Haystack" antenna system, 299–300
Heaviside, Sir Oliver, 4–5, 16, 314
HEOS satellite, 214
Hertz, Heinrich R., 3–8, 74, 82
  kilohertz (kHz); megahertz (MHz), 286
Hogan, John V. L., 61
Hollywood, 322–324
Holography, laser photography, 259, 261–263
Hoover, Herbert, 177
Houbolt, John C., 128–129
Hughes Aircraft, 160, 181, 185, 199, 231, 264
Hydrogen, 210, 230, 271, 302–303

Iconoscope, 68
IMP satellite, 212
Infrared, 85, 101, 193, 194, 197, 226, 251, 260, 276, 279
Institute of Electrical and Electronics Engineers, 49, 68, 74
Integrated electronics, 123, 144, 277–280
Intelsat satellites, 161, 187, 189–190
Intercosmos satellites, 216

International Business Machines Corp. (IBM), 168–169, 282, 289
International satellites, 212–214
International Telecommunications Satellite Consortium, 161
International Telecommunications Union, 154
International Telephone & Telegraph Co., 160, 161
Interplanetary communication, 165–168
Interstellar travel, 167–168
Ionization, radio blackout, 94
Ionosphere, 4–6, 82–84, 94, 148, 211–213, 272, 287–288, 291, 305
Italian satellites, 214
Ives, Herbert E., 64, 66

Jansky, Karl G., 296
Jastrow, Robert, 314
Jean, Sir James, 173
Jenkins, Charles F., 64–66
Jet Propulsion Laboratory, 156, 222, 225, 228, 229–233, 235, 237, 245, 254
Jodrell Bank Observatory, 242, 297, 310
Johns Hopkins University, 159, 204
Johnson, Lyndon B., 86, 105, 106, 117, 126, 163–164, 173, 187, 233, 327
Johnston, Eric, 322
Jupiter, 98, 247, 309–310

Kaempffert, Waldemar, 84
Kansas City Star, 65
Kappel, Frederick R., 174
Kazakhstan, Baikonour Space Center, 93, 121
KDKA, 34–35, 48
Kell, R. D., 64
Kelley, Thomas J., 129
Kelvin, Lord, 7, 16
Kemp, George S., 11
Kennedy, John F., 70, 88–89, 90, 93, 96, 112, 159–163, 177, 184, 186–187
Kennelly, Arthur E., 5
Kennelly-Heaviside layer, 4–5
Kershner, Richard B., 159
Khrunov, Yevgeny K., 121
Kinescope, 68
Kitt Peak National Observatory, 236, 306

Klystron, 75
Komarov, Vladimir M., 98, 111
Korad Corp., 248
Korn, Arthur, 55–56

Langmuir, Irving, 272
Lani Bird satellite, 189
Lasecon, 279
Laser, history and development, 246–268
    applications, 85, 101, 210, 252–263, 280, 283
    materials that lase, 263–267
    ranging, retro-reflector (LRRR), 145
    types of, 247–254, 263–267
Laserfax, 256
Lavoisier, Antoine, 302
Lebedev Institute of Physics, 311
Leiden Observatory, 310
Leith, Emmett N., 262
Leonov, Aleksei A., 99
Light, relation to radio, 1–4, 85; see also Coherent light
Light-year, 165–167, 312–313
Lincoln Laboratory, see Massachusetts Institute of Technology
Lindbergh, Charles A., 26, 95, 112
Liquid crystals, 283–284
Lodge, Sir Oliver, 11, 16, 17
Lofti satellite, 200
Logwood, Charles V., 22
Loomis, Mahlon, 16
Loran, 204, 206
Lovell, James, Jr., 101, 110, 122–127, 149
Luna spacecraft, 240–243
Lunar module (LM), 118, 127–129, 137–141
Lunar Orbiters, 236–240

MADA, 292
Magnetometer, 148, 209, 225
Magnetosphere, 203, 304
Magnetron, 75, 328
Maier, L. C., Jr., 277
Maiman, Theodore H., 248
Manned Spacecraft Center, see Mission Control Center
Marconi, Guglielmo, 6–18, 47, 48, 67, 74, 83–84, 115, 315, 317
Marconi Wireless Telegraph Co., 12, 17, 33, 319–320
Mariner spacecraft, 224–231, 329

Mars, 85, 97–98, 116, 126, 166
Mariner probes, 226–230
Soviet probes, 243, 310
observations by University of Michigan, 307
future explorations, 329
Martin Marietta Corp., 291–292
Maser, 246–252; see also Laser
Maser Optics, Inc., 252
Maser-surgery (laser), 252–253, 258
Massachusetts Institute of Technology, 75, 157, 181, 255, 266–267, 276, 299–300, 313
Maxwell, James Clerk, 1–3, 7, 17
McCormack, James, 161
McDivitt, James A., 100, 127–128
McNamee, Graham, 41, 43
Medical electronics, 87, 92, 100, 108, 114, 252–254, 258
Meghreblain, Robert, 226
Mercury Project, 86–96, 111
Meteoroids, 124, 145, 200, 209–210
Metropolitan Opera, 31, 33, 42
Michelson, A. A., 256–257
Microwaves, 9, 18, 27, 74–75, 85, 247–248, 255, 270, 271, 286, 288–289
Midas satellite, 199–200
Military satellites, 180–182
Miller, George P., 115
Mills, Bernard Y., 311
Millstone Hill Field Station, 299–300
Miniaturization, 81–82, 119, 123, 144, 152–153, 276, 277–280, 284–285
Mission Control Center, 101, 105, 106, 108, 123, 131, 137, 140–141, 143, 149
Modulation, AM and FM, 49–53, 147, 322
Molniya satellites, 190
Molonglo Radio Observatory, 311
Moon, challenge of, 84, 89, 97–98, 112–115, 255
lunar module (LM), 118, 127–129, 137–141
Apollo journeys to, 130–135, 136–147, 147–149
scientific experiments on, 144–148
explored by robot spacecraft, 223–224, 231–243
Moon relay system, 157–159
Morley, E. W., 256–257
Motion pictures, relation to TV, 322–325

NAA, Arlington, Va., 32
NASA, 86–87, 113, 115, 116
Mercury Project, 86–96
Gemini Project, 99–112
Apollo Project, 117–149
relay satellites, 183–190
space explorations, 150–157, 221–245, 313
National Aeronautics & Space Administration, see NASA
National Aeronautics & Space Council, 159, 163
National Broadcasting Co., see NBC
National Radio Astronomy Observatory, 302, 305–307
Navigation satellites, 198–199, 204–208
NBC, 44, 47–48, 66, 69
NC-flying boats, 23–25
Neutron stars, 201
Newspapers, relation to radio, 44–47, 281–283
New York Times, 2–3, 45, 111, 154, 179, 317
Nikolayev, Andrian G., 93
Nimbus satellites, 197–198
Nipkow, Paul, 64–65
Nixon, Richard M., 139, 140–141, 143, 149
"Norge," dirigible, 25–26
Nottingham University, 3
Nuclear generators, see SNAP

Ochs, Adolph S., 45, 48
Oliphant, Mark L., 75
Omega navigational aid, 206
Onnes, H. K., 275–276
Optical maser, 248–250, 252; see also Laser
Orbiters, see Lunar Orbiters
Orbiting observatories, 200–204
Oscar satellites, 200
Otter Cliffs, Bar Harbor, Maine, 25, 179
Ovshinsky, Stanford R., 284

Pageos satellite, 208
Paget, G. W., 11
Paine, Thomas O., 113
Palomar Mountain Observatory, 296, 309
Patrick Air Force Base, 78–79
Pauling, Linus, 303
Pay-TV, 323–325

Pearl Harbor, radar, 78
Peary, Admiral Robert E., 26
Pegasus satellite, 200
Perkin-Elmer Corp., 260
Phillips, Jack, 14
Phonograph, 31–33, 36, 42, 320–321
Picturephones, 327
Pierce, George, 16
"Pioneer" satellites, 150, 191–193
Planck, Max, 246
Plasma physics, 272–274
Polyot satellites, 243–244
Popoff, Alexander S., 16
Popovich, Pavel R., 93
Preece, Sir William, 9, 10, 16
Presidents on radio and TV, 41, 68, 70, 88–89, 140–141, 149, 187, 188, 327
Proton spacecraft, 216–217
Pulsar, 302
Purcell, Edward M., 167–168, 271

Quasar, 301–302
Queen Elizabeth 2, (QE2), 206

"R-34" dirigible, 25
RACEP, and RADA, 291–292
Radar, history and development, 73–85, 327–328
    Space Age applications, 86–149, 279
    use by Apollo 11 and 12, 141, 146, 149
    role in radio astronomy, 245, 295–315
    laser-radar, 256, 265
Radio, evolution of, 1–27, 316–321
    Space Age applications, 86–111
    on Apollo moon-missions, 118–149
    use by exploring spacecraft, 150–173, 191–219
    communication relay satellites, 174–190
    robot communicators' use of, 221–245
    astronomy, 200–204, 211–212, 295–315
    advances in the spectrum, 286–294
    See also Wireless Telegraphy, Radiotelephone and Radio Broadcasting
Radio astronomy, 200–204, 211–212, 221–224, 271, 295–315
Radio Broadcasting, history and development, 28–54, 321–322

Radio Club of America, 21
Radio-control, 18–20, 78–79, 170, 224, 231–236
Radio Corporation of America, see RCA
Radio echoes, 74–76, 82–83
Radio "music box," 32–33, 36, 41
Radiophoto, history and development, 55–62
Radio-radar telescopes, 295–315
Radiotelephone (radiophone), see Radio Broadcasting
Radome, 173
Raman, C. V., 264
Ranger, Richard H., 56–58, 61
Ranger spacecraft, 222–224
Raytheon Company, 255
RCA, 33, 36, 47–48, 60–61, 66, 68–69, 70–72, 77–78, 119, 146, 160, 161, 183, 257–260, 281–282, 318, 319–320
RCA Laboratories, 71, 256, 274–275, 279–280, 283–284; see also RCA
Reinartz, John L., 22
"Relay" satellite, 183–185
Republic, S.S., 13
Righi, Augusto, 6, 16
Rignoux, M. Georges, 65
Robot exploring communicators, 221–245
Rodd, Herbert C., 23–25
Roosevelt, Franklin D., 41, 46, 68, 327

Sagittarius, 312–313
Sanabria, Ulisses A., 64
Sarnoff, David, 15, 32–33, 270–271
Satellites, 152, 154–157, 172–173
    for exploration, 150, 191–220, 221–245
    for communications, 174–190
    for navigation, 204–208
Saturn rocket, 117, 118, 122, 127, 137
Scanning disk, 64–66, 67, 130, 144
Schawlow, Arthur L., 248
Schirra, Walter M., Jr., 93–94, 101, 118
Schnell, Fred H., 22
Schweickart, Russell L., 127–128
Score satellite, 151
Scott, David R., 105, 127–130
Secor satellite, 208
Semiconductors, 53–54, 81, 251, 276, 284

SERT, electronic rocket, 274
Shatalov, Vladimir A., 121
Shepard, Alan B., Jr., 86–87, 149
Shockley, William, 53
Shortwaves, 3–6, 21–22, 48, 320
Slaby, Adolphus, 16, 64
Slomar satellite, 219
Smithsonian Astrophysical Observatory, 210, 261, 305
SNAP nuclear generator, 148, 198, 207
Solar cells, 150–151, 175, 193, 197, 231
Solar winds, 148, 171, 200, 203, 212, 226
Solid-state physics, 53–54, 81, 152–153, 247, 251–252, 266–268, 289; see also Semiconductors
SOS, 13, 14
Southworth, George C., 176
Soviet Union, 86, 98–99, 111, 150, 162–164, 172–173, 214–218, 240–245, 310, 311
Soyuz spaceships, 111, 120–121
Spa-Sur, 170
Space Age communications, radio, TV, radar, 86–149, 150–173
national objective, 159–161
communications satellites, 174–190
exploring and weather satellites, 191–200, 209–218
navigation satellites, 204–208
orbiting observatories, 201–204
robot spacecraft, 221–245
Space exploration, value of, 89, 97–98, 113–115, 280, 329–330
immediacy in, 133
law and regulations, 161–165
Apollo moonships, 117–149
exploring satellites, 150, 191–220, 221–245
surveying the universe, 295–315
Space, launches in first decade, 172–173
Space law, 154, 161–165, 179
Space optics, 260–261
Space research, 113–114, 126–127, 134–135, 157–159, 191–219, 260–261, 267–268, 280, 283, 286–289, 329–330
Space treaty, UN, 161–164
Spectrum, 1–8, 286–294
Sperry Gyroscope Co., 256
Sputnik, satellites, 150

Stafford, Thomas P., 101, 106, 107, 130–132
Stanford University, 309
Steady-state theory, 300–301, 313–314
Steinmetz, Charles P., 2
Stereo FM, 52
Stevenson, Adlai, 162
Stormer, Carl, 82–83
Stormfinder radar, 80
Superheterodyne, 41, 52
Supreme Court, see U.S. Supreme Court
Surveyor spacecraft, 148, 231–236
Sylvania, laser, 254
Synchronous satellites, 155–156, 185–190
Syncom satellites, 185–188

Tacomsat, 181
Tacsatcom, 181–182
Taylor, A. Hoyt, 77, 82
Telemetry, 90, 91–92, 95, 97, 100, 108, 135, 137, 146, 147, 294
Telephones, 140, 159–160, 180, 183, 318, 327
Telescopes, 170, 201–202, 203, 211, 295–297, 304–314
Television, history and development, 63–72, 320–323, 325–327
Space Age applications, 86–111, 270, 279–280
Apollo Project's use of, 119–149
telecast of earth and moon, 121–127
first color TV from space, 130–133
first TV from moon, 138–142, 144
TV use by Apollo 12, 147–149
Telex, 58–59
Telstar satellite, 94–97, 174–179, 184
Tereshkova, Valentina V., 96–97
Tesla, Nikola, 16, 17, 18, 85, 315
Thaler, William J., 255
Thermo-electricity, 148, 276, 279, 330
Thomson, J. J., 9, 82
Timers, electronic-electrochemical, 184, 277
Times (of London), 48
Tiros satellites, 193–196
Titan rocket, 180–182
Titanic S.S., 13–15
Titov, Gherman S., 90
Tizard, Sir Henry, 75
Tomiyasu, Kiyo, 255–256
Tonks, Lewis, 272

Townes, Charles H., 246–248, 265
Transistors, 53–54, 81, 278, 284
Transit satellites, 204–208
Troposcatter, 290–291
Tsiolkovsky, Konstantin E., 99
Turkevich, Anthony, 235

Ultrafax, 60–61
United Nations, 115–116, 154, 161–164
U.S. Air Force, 77, 171, 308
  Defense Command, 172–173, 180–181, 202–203
U.S. Army Signal Corps, 77–78, 84, 151, 257
U.S. Army Signal Research & Development Laboratory, 150
U.S. Department of Defense, 87, 173, 180–182, 240, 293
U.S. Naval Research Laboratory, 82, 247; see also U.S. Navy
U.S. Navy, 22, 23–25, 32, 77, 82, 157–159, 179, 204–206, 319
U.S. Senate, 160–161, 164
U.S. Supreme Court, 17, 31, 72
U.S. Weather Bureau, 80–81, 192, 196
University of California Observatory, 309
University of Illinois Observatory, 307
University of Michigan, 262, 297, 307
University of Texas Observatory, 308–309
Upatnieks, Juris, 262

Van Allen, James Alfred, 271
Van Allen belts, 150, 185, 207, 210, 215, 225, 230, 271, 309
Van Anda, Carr V., 45, 83
Vandenberg Air Force Base, 157, 196, 200, 207, 208
Vanguard satellites, 151
Varian, Russell and Sigurd, 75
Vela "watchdog" satellites, 182
Venera satellites, 244–245
Venus, Mariner probe, 224–226, 229–230
  Venera probe, 244–245
  radar probes, 297–298
  observations by University of Michigan, 307
Verne, Jules, 113, 137, 139

Victor Talking Machine Co., 36, 42, 320–321
Videocomp, 281–282
Videoscan, 278
Viking spacecraft, 329
Volynov, Boris V., 121
Voskhod spacecraft, 98–99
Vostok spacecraft, 86, 90, 93, 96–97

Walker, Joseph A., 26–27
Wallops Island, 195, 213–214, 274
Watson Research Center, Thomas J. (IBM), 267, 289
Watson, Thomas J., Jr., 169
Watson-Watt, Sir Robert, 76–77
Waveguides, 176, 249, 279
Weather-radar, 80–81
Weather satellites, 193–198
Webb, James E., 113–114, 117
Welch, Leo D., 161
Wells, H. G., 43, 45
Western Electric Co., 39, 290
Western Union, 130, 161, 288, 293–294
Westinghouse Electric Co., 33–35, 36, 130, 144
Wheeler, Harold A., 176
Whipple, Fred L., 210
White, Edward H., 100, 111
White, General Thomas D., 171
"White Alice," Alaskan network, 290–291
Wiesner, Jerome B., 115
Wireless telegraphy, history and development, 1–27, 316–320, 328
Wolfe-Barry, Sir John, 317
World's Fair, New York, 1939 and 1964, 68–69, 327

X-rays, 85, 182, 201–203, 211–212
X-15 rocket plane, 26, 27

Yegorov, Boris B., 98
Yeliseyev, Aleksei S., 121
Young, Charles, 59
Young, John W., 99–100, 107–108, 130–133
Young, Leo C., 77
Young, Owen D., 59
Young, Thomas, 261

Zeppelin, radio, 25
Zond satellites, 217–218
Zworykin, Vladimir K., 64, 67–68

72 73 74 10 9 8 7 6 5 4 3 2